OX FORD HANDBOOKS IN EMI
Series Editors R. N. Illingworth, C. E. F

This series has already established itself as the essential reference series for staff in A & E departments.

Each book begins with an introduction to the topic, including epidemiology where appropriate. The clinical presentation and the immediate practical management of common conditions are described in detail, enabling the casualty officer or nurse to deal with the problem on the spot. Where appropriate a specific course of action is recommended for each situation and alternatives discussed. Information is clearly laid out and easy to find—important for situations where swift action may be vital.

Details on when, how, and to whom to refer patients are covered, as well as the information required at referral, and what this information is used for. The management of the patient after referral to a specialist is also outlined.

The next of each book is supplemented with checklists, key points, clear diagrams illustrating practical procedures, and recommendations for further reading.

The Oxford Handbooks in Emergency Medicine are an invaluable resource for every member of the A & E team, written and edited by clinicians at the sharp end.

Acute Medical Emergencies

Ursula Guly
Consultant in Accident and Emergency Medicine,
Dundee Royal Infirmary, Dundee

and

Drew Richardson
Director of Emergency Medicine
Princess Alexandra Hospital, Brisbane

Oxford • New York • Tokyo
OXFORD UNIVERSITY PRESS
1996

Oxford University Press, Walton Street, Oxford OX2 6DP

Oxford New York
Athens Auckland Bangkok Bombay
Calcutta Cape Town Dar es Salaam Delhi
Florence Hong Kong Istanbul Karachi
Kuala Lumpur Madras Madrid Melbourne
Mexico City Nairobi Paris Singapore
Taipei Tokyo Toronto
and associated companies in
Berlin Ibadan

Oxford is a trade mark of Oxford University Press

Published in the United States
by Oxford University Press Inc., New York

A catalogue record for this book is available from the British Library

Library of Congress Cataloging in Publication Data
Guly, Ursula.
Acute medical emergencies / Ursula Guly and Drew Richardson.
p. cm. — (Oxford handbooks in emergency medicine ; 14)
Includes index.
1. Medical emergencies—Handbooks, manuals, etc. I. Richardson,
Drew. II. Title. III. Series.
[DNLM: 1. Acute Disease—handbooks. 2. Emergencies—handbooks.
3. Critical Care—handbooks. WB 39 098 v. 14 1996]
RC86.8.G85 1996 616.02′5—dc20 DNLM/DLC 95–45956
ISBN 0 19 262559 4 (Hbk)
ISBN 0 19 262558 6 (Pbk)

Typeset by Footnote Graphics, Warminster, Wilts
Printed in Great Britain by
Biddles Ltd, Guildford

Preface

The majority of critically ill patients presenting to an Accident and Emergency department will be suffering from medical conditions. When dealing with such patients it is essential that appropriate treatment is instituted rapidly to prevent further deterioration in the patients' condition. Such patients are often unable to give a history and have no immediately obvious diagnosis, but frequently the medical staff responsible for the initial management of these severely ill patients are among the most experienced doctors working in the hospital.

At present, undergraduate and much postgraduate teaching is based around the management of patients with specific diagnoses, rather than what to do as the only doctor faced with a moribund patient in the Accident and Emergency department, when no diagnosis is known.

This book does cover the management of specific medical emergencies, but also aims to give a structure to the initial management of severely ill patients. This follows the Airway, Breathing, Circulation system advocated in cardiac and trauma life support courses. The advantage of using the ABC approach in all critically ill patients is that reversal of airway obstruction, hypoxia, and shock is not delayed while the diagnosis is made.

Inexperienced medical staff should not be expected to undertake the resuscitation and ongoing management of critically ill patients unaided, but this book aims to enable them to undertake the resuscitation and management until senior help arrives.

Dundee U.M.G
Brisbane D.R
January 1996

v

Contents

CHAPTER 1

How to approach the ill patient

- **Airways, Breathing, Circulation (ABC) 1** **Before the patient arrives 1** **Resuscitation 2** **After resuscitation 3**

Introduction

At medical school we learned that the standard way to approach a patient was the sequence of history, examination, investigation, then treatment. In the critically ill patient this order must change because the treatment must often precede the diagnosis. Fortunately, the initial assessment and management of the ill patient is identical regardless of the underlying condition.

Airway, Breathing, Circulation (ABC)

By following the ABC procedure during the assessment and treatment of every ill patient you will avoid missing any immediately life-threatening conditions. If at any point you feel you are out of your depth call for senior help. A doctor who realizes his/her limitations is a safe doctor and the type we would wish for ourselves and our own families.

Before the patient arrives

You may get warning from the general practitioner (GP) or ambulance control that a critically ill patient is expected. If you do not think you have the manpower or expertise to manage, call for senior help at once. Ensure all the equipment you need is available and functioning correctly. Check through what you might need for:

A suction, oropharyngeal airways, tracheal tubes, and a laryngoscope;

B oxygen, high concentration oxygen mask, and self-inflating bag with reservoir, valve, and mask;

C electrocardiogram (ECG) monitor/defibrillator, various sizes of IV cannulae, a giving set that is run through with fluid and drugs. Ensure that you and all the team are wearing appropriate protective clothing with eye protection and gloves—assume every patient is high risk.

Resuscitation

When the patient arrives do not allow yourself to be distracted from carrying out the ABC procedure. Tell the ambulance crew you will need to talk to them soon but do not stop to listen to the history while the patient's airway obstructs.

A Ensure the airway is clear. If it is occluded you must not move on until it is patent. Almost always simple airway manoeuvres, such as head-tilt/chin-lift and jaw-thrust, are all that is required to open the airway and allow ventilation.

B Ensure the patient is ventilating with high concentration oxygen. A mask with a reservoir connected to high-flow oxygen is ideal for the spontaneously ventilating patient. If the patient is apnoeic or has a very low respiratory rate ventilate the patient with a self-inflating bag with reservoir, valve, and facemask connected to high-flow oxygen. Use a two-person technique, one holding the mask and maintaining the airway and the other squeezing the bag. Listen to both sides of the chest to ensure there is adequate air entry and to check there are no signs of a tension pneumothorax.

C Check there is a pulse. Assess the peripheral perfusion by measuring the capillary refill time. Check the heart rate. Secure IV access in a large relatively proximal vein, (e.g. in the antecubital fossa). Measure the blood pressure and connect the patient to a pulse oximeter and ECG monitor.

After resuscitation

History Once ABC have been assessed and resuscitation started you can safely take a brief history excluding details not immediately relevant. Often no history can be obtained from the patient so remember other sources such as ambulance staff, police, relatives, neighbours, home-helps, and wardens. You may need to phone them.

Examination Examine the whole patient, what you forget may be overlooked later. Where the history is unclear remember to look for signs of injury—stab wounds and rib fractures can also cause breathlessness and chest pain. If at any point the patient deteriorates reassess ABC.

By the time you have taken a brief history and examined the patient you should have an idea of a differential diagnosis and be able to formulate a logical management plan. By assessing and treating ABC first you have already started the patient's treatment and kept them alive for a diagnosis to be made.

Once the patient is stable, talk to the relatives. They should have been kept informed of the patient's condition by another member of staff but as team leader you should see them to explain the situation and answer any questions.

CHAPTER 2

Cardiac arrest

Basic life support and advanced airway techniques

This section is aimed at staff in the *accident and emergency (A&E) department and so differs slightly from the Resuscitation Council guidelines for basic life support (BLS) in the community. The priorities of ABC remain identical.*

The role of BLS in A&E   This aims to maintain cerebral oxygenation until the underlying cause of the arrest can be reversed. BLS only provides about 20 to 30 per cent of normal cerebral and myocardial blood flow, and since successful defibrillation from ventricular fibrillation (VF) becomes less likely the longer it is delayed, nothing, not even BLS should delay defibrillation for a patient in VF. Apart from defibrillation BLS should not be interrupted. It is easy to assume someone else is remembering BLS but it is *your* job to ensure it is remembered.

Call for help   SHOUT for help. If no one else has done so call the cardiac arrest team. You must know the emergency number and state exactly where you are.

Open the airway   Look in the mouth and remove any foreign body (e.g. food or loose dentures). Remove liquid by suction with a rigid Yankauer catheter.

4

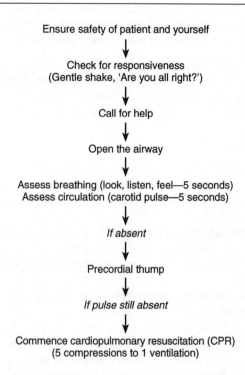

Ensure safety of patient and yourself
↓
Check for responsiveness
(Gentle shake, 'Are you all right?')
↓
Call for help
↓
Open the airway
↓
Assess breathing (look, listen, feel—5 seconds)
Assess circulation (carotid pulse—5 seconds)
↓
If absent
↓
Precordial thump
↓
If pulse still absent
↓
Commence cardiopulmonary resuscitation (CPR)
(5 compressions to 1 ventilation)

Simultaneously arrange for the patient to be connected to an ECG monitor/defibrillator

Fig 2.1 • Summary of basic life support in the A&E department.

Head-tilt/chin-lift. Put one hand on the hairline and tilt the head into extension then put two fingertips under the point of the chin and lift. If this does not open the airway use the jaw-thrust manoeuvre. Put your fingers at the angles of the jaw and lift it forwards.

Assess for breathing and circulation
• Put your cheek over the patient's mouth, **look** for chest movement, **listen** for breath sounds, **feel** for expired air for 5 seconds before diagnosing apnoea.
• Feel for a carotid pulse for 5 seconds before deciding it is absent.

The diagnosis of cardiac arrest is a clinical one based on unconsciousness and the absence of central pulses. Once diagnosed the circulation must be restored by treating ABC

Precordial thump In a witnessed arrest a precordial thump may terminate ventricular fibrillation (VF) or ventricular tachycardia (VT). It is delivered with a clenched fist from a height of 15–30 cm at the same site as chest compressions.

Ventilation

It is essential to ventilate the patient with as high a concentration of oxygen as possible. It is often tempting to rush into intubation but this takes time and the patient must be oxygenated first.

Expired air resuscitation only delivers 16 per cent oxygen so its use is restricted to situations where no equipment is available. It is well described in other texts but is rarely used in hospital.

Mouth to mask ventilation uses expired air through a unidirectional valve which is then supplemented by oxygen. The airway is kept open by a head-tilt/jaw-thrust, the thumbs keeping the mask sealed on to the face. It has aesthetic advantages over mouth to mouth and it is easily performed by one person. Become trained in its use before using it.

The oropharyngeal (or Guedel) airway This is designed to keep the tongue forward, off the posterior pharyngeal wall, thereby keeping the airway open and aiding ventilation. The correct size is found by holding the airway to the patient's face. The length of the correct airway equals the distance between the incisor teeth and the angle of the jaw.

The usual way to insert an oropharyngeal airway is to put it in the mouth upside down and then rotate it by 180 degrees as it passes into the oropharynx. Alternatively, the tongue may be held down by a tongue depressor or a laryngoscope blade and the airway slid over it.

Bag-valve-mask technique This is a method of ventilation which looks easy but is not. The airway must be held open, with a good seal between the mask and the patient's face for the patient to be ventilated. This is difficult to do with one hand on the mask while squeezing the bag with the other hand. It is all too easy to fail to ventilate the patient at all. It is best to use a two-person technique where one person holds the mask on the patient's face with two hands, achieving a good seal and keeping the airway open with a jaw thrust while the other squeezes the bag—85 per cent oxygen can be given by connecting high-flow oxygen to the system.

Laryngeal mask airway   The airway may be maintained open following cardiac arrest using a laryngeal mask airway. This requires less training than tracheal intubation but has the disadvantage that the lower airway is not protected against aspiration of gastric contents. This technique must be learned under supervision before it is undertaken.

Tracheal intubation   This has the advantage of both opening and protecting the airway. As stated above it is never the first airway manoeuvre but is performed only when the patient has been pre-oxygenated by another method (For more information see Chapter 16).

Chest compression

Place the side of the heel of one hand over the sternum, starting two finger-breadths from the xiphisternum in the midline. Put your other hand on top of it, lifting your fingers to keep them off the ribs.

Lock your elbows straight and with your shoulders directly above your hands press vertically downwards 4–5 cm, releasing the pressure completely in the relaxation phase. To achieve effective chest compression with a patient on an A&E trolley, it is usually necessary to stand on a footstool. The compression rate is 80 per minute.

Combining ventilation and chest compression

Single rescuer CPR: 15 compressions to 2 ventilations. Compression rate 80/min. Recheck your hand position each cycle.

Two-rescuer CPR: 5 compression to 1 ventilation. Compression rate 80/min.

The compressions and ventilations should occur one after the other and not simultaneously. If one person counts the compressions out loud it makes the co-ordination easier.

Conclusion Basic life support must be learned and kept up to date by practical training using a mannikin. It cannot be learned from a book. To refresh your BLS skills contact your hospital's resuscitation training officer. If you feel you need training or retraining in advanced life support skills apply to attend an advanced life support course. Details of which centres near you organize these courses can be obtained from the Resuscitation Council (UK), 9 Fitzroy Square, London W1P 5AH.

Ventricular fibrillation (VF) and pulseless ventricular tachycardia (VT)

VF and pulseless VT are treated identically. The likelihood of successful defibrillation depends on the delay before defibril-

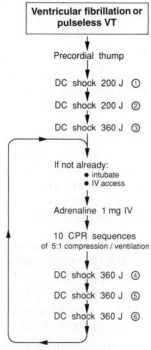

Fig 2.2 • Flowchart for the management of ventricular fibrillation/pulseless ventricular tachycardia.

lation, decreasing after only 90 seconds from the onset of VF/VT so nothing, not even BLS should delay defibrillation.

After each shock check the patient by feeling for a central pulse and check the rhythm shown on the monitor while recharging the defibrillator.

Modern defibrillators charge very rapidly so it should be possible to deliver each series of 3 shocks within 30 seconds without delaying to give cardiopulmonary resuscitation (CPR) between shocks. If you have an old, slow defibrillator or your team cannot deliver the 3 shocks rapidly give one cycle of CPR (i.e. 5 compressions and 1 ventilation) between shocks. Intubation and IV access should not be allowed to interrupt CPR for more than 30 seconds or delay further shocks.

If there is any doubt about the efficacy of the defibrillator change it; you may also wish to try altering the paddle positions to anteroposterior.

If, after 3 loops of the algorithm, the patient remains in VF other drugs may be considered, although they have not been shown to be of unequivocal benefit in this setting. The most common anti-arrhythmic drug used in lignocaine 100 mg IV. Bretylium tosylate or amiodarone are alternatives that could be used. Resuscitation should not be discontinued while the patient remains in VF or VT.

Asystole

Asystole is diagnosed by the absence of ventricular electrical activity on the ECG in a patient with a cardiac arrest. Isolated P waves may occur. The ECG is rarely a true 'straight line'. If this is seen the cause is almost certainly a disconnected lead.

Asystole carries a poor prognosis. If you are uncertain if the rhythm could be VF treat as for VF, with defibrillation.

Box 1.1 **In every case where you suspect asystole**

- Check the monitor is connected to the patient
- Check that the correct lead is selected on the monitor
- Turn up the gain on the monitor

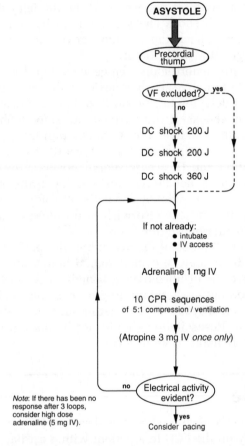

Fig 2.3 • Flowchart for the management of asystole.

Electromechanical dissociation

Electromechanical dissociation or EMD (sometimes called pulseless electrical activity or PEA) occurs when there is a rhythm on the cardiac monitor but the patient is pulseless. When it is due to myocardial dysfunction, following acute myocardial infarction (MI) it carries a very poor prognosis but there are some important treatable causes that must be considered in every case, while BLS is started.

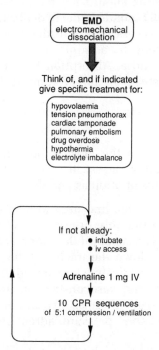

Fig 2.4 • Flowchart for the management of electromechanical dissociation.

There is no evidence that routine use of calcium is of benefit. Calcium chloride 10 ml 10 per cent IV is only indicated for EMD associated with hyperkalaemia, hypocalcaemia, or overdose of calcium-channel blockers.

Drugs in cardiac arrest

Route of administration

During CPR, it is preferable to give all drugs into a central vein so that they reach the heart more quickly. Until central venous access is achieved drugs should be given through a cannula in a large, relatively proximal peripheral vein, such as the external jugular or an antecubital fossa vein. Insertion of a central venous cannula should not be allowed to interrupt CPR for more than 30 seconds or delay defibrillation.

If venous access proves difficult, adrenaline, atropine, and lignocaine may be administered down the tracheal tube at twice the IV dose. Absorption from the lungs in a patient in cardiac arrest is poor and IV administration is preferable when possible.

Specific drugs

Oxygen This is the most important drug used during and following cardiac arrest. It should be given to all patients in as high a concentration as possible.

Adrenaline is the first drug to be given in all cardiac arrest rhythms and the only drug in the VF/pulseless VT and EMD algorithms. The standard dose for an adult is 1 mg IV, doubled if it is given down the tracheal tube. There is little evidence that high-dose adrenaline is of any additional benefit. Adrenaline causes peripheral vasoconstriction, thereby increasing aortic diastolic pressure. Because myocardial blood flow is dependent on aortic diastolic pressure, adrenaline improves myocardial perfusion.

No other inotrope has been shown to have an advantage over adrenaline in cardiac arrest.

Atropine is a parasympathetic blocking agent. It increases the rate of discharge of the sinus node, the conductivity, and spontaneous discharge rate of the AV node and also reverses vagally mediated suppression of myocardial function.

It appears on the asystole algorithm, after adrenaline, given once only on the first loop in a dose of 3 mg. The rationale behind its use in this setting is that the appearance on the cardiac monitor may be caused by a profound vagally mediated bradycardia such as occurs in the diving reflex. It is the first-line drug in bradycardia with a cardiac output.

Sodium bicarbonate Patients following cardiac arrest develop a mixed respiratory and metabolic acidosis and in the past sodium bicarbonate was given routinely to reverse the acidosis. However, the acidosis shifts the oxygen dissociation curve thereby increasing cellular oxygenation, so reversal can increase tissue hypoxia and sodium bicarbonate administration results in the release of carbon dioxide (CO_2), which readily passes into the cells, paradoxically increasing the intracellular acidosis. It is also hyperosmolar and provides the heart with a massive sodium load. Its use in adult cardiac arrest is now restricted to patients who are being effectively ventilated in whom BLS, defibrillation, and anti-arrhythmic drugs have all failed to restore the circulation. The dose is 1 mmol/kg (i.e. 1 ml/kg of 8.4 per cent solution). The most important aspect of acid–base management is effective ventilation, leading to CO_2 excretion.

Lignocaine is an anti-arrhythmic agent that can sometimes be effective in treating ventricular tachycardia. Because of this it was routinely used in the past for the treatment of resistant ventricular fibrillation. However, not only has lignocaine never been shown to be of benefit in this setting, but experimental evidence suggests it may make successful defibrillation less likely. It may be given in VF if the loops of the algorithm have not resulted in successful defibrillation.

Calcium Calcium is involved in the intracellular initiation of cardiac contraction. There is, however, no good evidence that giving additional calcium is of benefit in cardiac arrest and it may cause myocardial and cerebral damage in the presence of ischaemia. Because of this, calcium is indicated during cardiac arrest only in EMD with the specific indications of, hyperkalaemia, hypocalcaemia, or calcium-channel blocker toxicity. The dose is 10 ml 10 per cent calcium chloride IV.

Bretylium tosylate is a drug with complex effects on the adrenergic nervous system, and was initially marketed as an antihypertensive agent. It is effective against ventricular arrhythmias but because of its marked hypotensive effect it has rarely been used in patients with a cardiac output since the advent of modern anti-arrhythmics. It may be used in VF or pulseless VT following three unsuccessful loops of the algorithm. Because its anti-arrhythmic effect can be delayed, full resuscitation should be continued for at least 20 minutes after its administration. Despite anecdotal case reports there are no studies that show it to be superior to lignocaine in resistant VF.

When to stop resuscitation

When patients arrive in the A&E department in cardiac arrest, resuscitation should be started or continued until the history becomes clear. If resuscitation were discontinued in order to get a history from relatives or the ambulance crew hypoxic brain injury would occur.

Resuscitate first, ask questions during and after

Resuscitation should be stopped only when it becomes clear that no recovery is possible. This is usually either when a history becomes clear of the cardiac arrest being the terminal event of an untreatable disease, such as malignancy, or when the resuscitation has continued for so long, without any response, that hypoxic brain injury is inevitable. Unfortunately, even expert CPR provides suboptimal cardiac output and will not sustain normal cerebral oxygenation.

Box 2.2 **Indications for prolonged resuscitation**

- Poisoning with CNS depressant drugs (e.g. barbiturates)
- Hypothermia

In both these situations there is slowed cerebral metabolism and there may be intact neurological function despite prolonged CPR.

Importantly, profound hypothermia can have all the physical signs of cardiac arrest. Resuscitation should not be stopped while the core temperature is below 32 °C—remember the maxim: 'They're not dead 'til they're warm and dead.'

The decision to stop should be made by the team leader in consultation with the team, including the nursing staff. It rarely happens that a team member disagrees and wishes to continue the resuscitation. If this does occur, the team leader should continue but with a clear end point, such as stopping if the patient remains in asystole after another 5 minutes CPR.

Post-resuscitation care

The return of a palpable cardiac output is not the end of resuscitation, it is the beginning of a more challenging phase. Every patient requires:

A A clear protected airway.
B Adequate ventilation and high concentration oxygen.
C Good tissue perfusion, IV access, and cardiac monitoring.

Some patients, usually those who underwent rapid defibrillation from VF will be able to be transferred to the coronary care unit (CCU), self-ventilating; but others will require ventilation, invasive monitoring, and cardiovascular support in the intensive therapy unit (ITU).

Assessment

History Including previous medical history and drug history.

Examination Full examination, including looking for signs of injury resulting from CPR.

Investigations

– *Chest X-ray* Check for position of tracheal tube and central line and signs of injury.
– *ECG* Particularly examine rhythm and look for signs of ischaemia or infarction.
– *Electrolytes* The potassium level should be maintained between 4 and 5 mmol/l.
– *Glucose* There should be normoglycaemia
– *Arterial blood gases* These should be assessed in any unconscious patient to assess adequacy of ventilation and the severity of metabolic acidosis.

It is impossible to predict eventual neurological outcome on the basis of neurological examination immediately following cardiac arrest. The pupils may be fixed and dilated due to atropine or catecholamines given during the arrest. Even patients who are initially apnoeic and Glasgow Coma Scale 3 (see Chapter 6) may make a full recovery. Every patient should be referred for ITU treatment if they require it.

Treatment

Airway If a patient does not have an open airway which they can protect against aspiration they require tracheal intubation.

Breathing All patients require high-flow oxygen. Ventilation should be assessed and if the patient cannot maintain good oxygenation, clearing CO_2 normally with a normal pattern of respiration they will require assisted ventilation. Following CPR, gastric dilatation is common, particularly following mouth-to-mouth or mouth-to-nose ventilation. The distended stomach can impede ventilation and provoke regurgitation and subsequent aspiration. Therefore, a nasogastric tube should be passed.

Circulation All patients require IV access and cardiac monitoring for at least the first 36 hours following cardiac arrest.

1. *Rhythm.* The patient needs have a rhythm capable of providing adequate tissue perfusion. If the cardiac arrest was due to VF, prophylactic anti-arrhythmic treatment may be required to prevent its recurrence. Locally agreed protocols should be followed recognizing that such treatment may have adverse effects.

2. *Blood pressure/perfusion.* Patients may have normal perfusion immediately following a cardiac arrest but myocardial dysfunction, associated with hypotension and poor perfusion is common because of prolonged myocardial ischaemia. It may be tempting to start inotropes in the A&E department to restore a normal blood pressure but doing this without a measure of cardiac function can be dangerous. At the very least, the central venous pressure, but preferably the left atrial ('wedge' or PAWP) pressure and cardiac output should be known. In that way the appropriate inotrope can be selected depending on the patient's cardiovascular status.

The best measure of tissue perfusion is urine output. All but haemodynamically normal, fully conscious patients should have a urethral catheter and hourly urine output measurements.

3. *Thrombolysis.* If the 12-lead ECG shows evidence of an acute myocardial infarction consideration should be given to thrombolysis if there is no contraindication. Prolonged 'traumatic' CPR is considered a relative contraindication but a brief period of CPR is not.

Neurological function Hypoxic brain injury is common as a result of cardiac arrest and may continue afterwards if the patient is not managed correctly. Adequate oxygenation and perfusion are essential and seizures, which increase the brain's metabolic rate, should be controlled. Although many other treatments, such as steroids, have been tried they are not of benefit and should not be used.

> **Remember to talk to the relatives**

Further reading

1. Skinner, D. V. and Vincent, R. (1996). *Cardiopulmonary resuscitation*, (2nd edn). Oxford University Press.
2. Handley, A. J. and Swain, A. (ed.) (1994). *Advanced life support manual*. Resuscitation Council (UK).
3. Rosenthal, R. (1992). Adult cardiopulmonary arrest. In *Emergency medicine: concepts and clinical practice*, (ed. P. Rosen), pp 106–44 Mosby-Year Books, St. Louis, MO.
4. European Resuscitation Council Basic Life Support Working Group (1993). Guidelines for basic life support. *British Medical Journal*, **306**, 1587–9.
5. Chamberlain, D., Bossaert, L., *et al.* (1992). Guidelines for advanced life support. A statement by the Advanced Life Support Working Party of the European Resuscitation Council. *Resuscitation*, **24**, 111–21.

CHAPTER 3

Cardiac emergencies

Assessing the patient with chest pain

Rapid assessment and resuscitation

1. Assess airway, Breathing, and Circulation. The patient may need resuscitation.

2. If the patient has had a myocardial infarction you do not want thrombolysis to be delayed unnecessarily, so unless there is an obvious non-cardiac cause, such as trauma, treat the pain as cardiac until proven otherwise. Give high-flow oxygen, obtain IV access and connect to a cardiac monitor. Order a 12-lead ECG.

Types of pain

This section aims to cover some of the more common causes of chest pain.

- *'Cardiac' pain*   The characteristics of cardiac chest pain are well known; severe central chest pain radiating to the jaw or left arm, described as crushing, or tight, associated with nausea, sweating, and fear. It is important to remember that the ECG may be normal following acute myocardial infarction (MI). Other causes of 'typical' cardiac pain include massive pulmonary embolus, aortic dissection, and cocaine.

- *Pleuritic pain* There are many differential diagnoses for pleuritic pain, the most common being minor trauma, but it is important to consider pneumothorax, pulmonary embolus, pneumonia, and pericarditis in all patients.
- *Atypical pain* The pain of MI can be 'atypical' with a normal ECG, and ischaemic heart disease can occur in people at any age. If you cannot make a positive diagnosis of a non-cardiac cause of the chest pain and think it could be cardiac refer the patient for admission. It is better for a patient to be admitted unnecessarily than to develop VF on the way home.

Analgesia

In your enthusiasm to make the correct diagnosis remember to treat the pain. If ischaemic cardiac pain is not rapidly relieved by nitrates, or if the patient has severe pleuritic pain with shallow breathing, give I.V. opioids with an anti-emetic, titrated to relieve the pain without depressing the conscious level.

Acute myocardial infarction

Initial assessment

When you first suspect a patient could have had an MI prepare for the worst:

- Think ABC.
- Put the patient on high-flow oxygen.
- Obtain IV access and connect to a cardiac monitor.
- Arrange for a 12-lead ECG and analgesia.
- Brief history and examination.

Time to thrombolysis is important in reducing myocardial injury and death so find out the local protocol to 'fast track' patients with possible MI to allow the earliest possible treatment.

History

The classical history in acute MI is of central tight, crushing chest pain radiating to the jaw or left arm, accompanied by sweating and nausea or vomiting. In patients who have angina, pain not relieved by GTN or that persists for over 10–20 minutes must be assumed to be due to MI until proved otherwise. Many patients do not present with a classical history. The patient may be unable to clearly describe their pain; the pain may be restricted to the arm or jaw; and in the elderly or diabetics there may be no pain, thus in ill patients consider MI.

Have a low threshold for suspecting MI and admitting the patient.

Investigation

12-lead ECG Remember that a normal ECG does not exclude MI so if you suspect the diagnosis clinically, refer the patient to CCU for admission.

Chest X-ray This gives an indication of heart size and the presence of left ventricular failure or other diagnoses such as aortic dissection.

Urea and electrolytes and glucose Hypokalaemia increases the risk of arrhythmias and must be corrected. Hyperglycaemia worsens the outcome from MI and is treated with an insulin infusion.

Cardiac enzymes There are kits for 'bedside' analysis of cardiac enzymes, which give results very rapidly, but these are not widely available. In most hospitals, it takes over an hour to obtain the results of the cardiac enzyme and thrombolysis must not be delayed until they are available.

Full blood count Severe anaemia increases myocardial work and may need urgent correction.

Treatment

Oxygen All patients should receive high-flow oxygen. Even if they have normal oxygen saturation their myocardium is ischaemic at the edges of the infarcted area. The only exceptions are the small proportion of patients with chronic lung disease known to retain CO_2

Nitrates Patients with ischaemic cardiac pain should be given nitrates sublingually unless they are hypotensive or have already had repeated doses.

Analgesia If the patient is in pain do not delay giving analgesia until you have completed an exhaustive history and a meticulous examination. Perform a brief history and examination and give analgesia early. This is not only humane but it reduces catecholamine levels and hence myocardial oxygen demand. Opioids should be given IV by titration to relieve pain but not reduce the level of consciousness. An antiemetic, such as metoclopramide, should also be given IV. Do not give intramuscular (IM) injections as drugs will not be absorbed, cardiac enzyme analysis will be difficult, and subsequent thrombolysis will cause haematomas.

Aspirin given alone is as effective as thrombolysis alone and is synergistic with it in reducing mortality and infarct size. The dose is 150 mg orally given as soon as possible. the only absolute contraindications are known allergy to aspirin, or asthma.

Thrombolysis

Multicentre double blind controlled trials in the 1980s proved that the mortality from MI could be halved by the early administration of thrombolytics. It is essential that thrombolysis is not delayed or forgotten.

Patients with a proven MI within the last 12 hours and no contraindications should receive thrombolysis as soon as possible. Studies have shown most benefit in patients with ST elevation or acute left bundle branch block but no improvement in survival in patients with a normal ECG or ST depression. If there is a relative contraindication to thrombolysis the potential harm must be weighed against the potential benefit. Discuss these patients with the cardiology team prior to starting thrombolysis.

Preparation for thrombolysis Take a history, including allergies, previous thrombolysis, and other contraindications, Perform an examination, looking for conditions that may mimic MI such as aortic dissection. Obtain secure IV access and put the patient on high-flow oxygen. Perform a 12-lead ECG and connect the patient to a cardiac monitor in an area with full resuscitation facilities.

There is little evidence that any thrombolytic is superior to any other. Usual practice is to use streptokinase, except when it has been used more than 5 days and less than 12 months previously, when tissue plasminogen activator (rt-PA) is used.

Box 3.1 **Absolute contraindications to thrombolysis**

- Recent surgery, major trauma, or bleeding (including aortic dissection)
- Pregnancy and puerperium
- Recent active peptic ulcer disease, known oesophageal varices
- Cerebrovascular disease, particularly recent cardiovascular accident or residual disability
- Severe hypertension (systolic > 200, diastolic > 115 mmHg)
- Uncontrollable clotting disorders, acute pancreatitis

For streptokinase/anistreplase, previous allergy to, or treatment with either drug between 5 days and 12 months previously. Use rt-PA instead.

Proliferative diabetic retinopathy is now thought to be a relative and not an absolute contraindication to thrombolysis.

Box 3.2 **Streptokinase and rt-PA dosage**

- *Streptokinase.* 1.5 million IU in 50–200 ml 5% dextrose or normal saline by IV infusion over 30–60 minutes.
- *rt-PA.* Total dose is 100 mg (1.5 mg/kg for patients < 67 kg) given over 3 hours. 10% of the total dose as IV bolus over 1–2 minutes then 50% of the total dose by infusion over 1 hour then 40% of the total dose by infusion over 2 hours.

Side-effects

Arrhythmias Reperfusion arrhythmias may be restricted to frequent ventricular extrasystoles but VF and VT also occur. Therefore, at all times including during transfer, the patient must be accompanies by full resuscitation equipment, a defibrillator, and staff trained to use them.

Hypotension Temporarily stop the infusion and when the blood pressure is normal restart at a slower rate.

Allergy/Anaphylaxis This is treated in the standard way with oxygen, fluid, and adrenaline, depending on severity (see Chapter 12)

Haemorrhage Major bleeding occurs in less than 0.5 per cent of patients. Stop the thrombolytic immediately and start resuscitation. Discuss with haematologists using an antifibrinolytic agent such as tranexamic acid.

Beta blockers

There is good evidence for improved outcome in patients with an acute MI discharged home on beta blockers. These would usually be started only after the patient has arrived in the CCU, following thrombolysis.

Cardiogenic shock

Myocardial infarction may be complicated by cardiogenic shock with hypotension and hypoperfusion. This is usually due to poor ventricular function and may require inotropic support. It is important that inotropes are not started without invasive monitoring, central venous pressure at the very least, and preferably a pulmonary artery catheter as some of these patients have a component of hypovolaemia contributing to their haemodynamic disturbance and this must be corrected before inotropic support is started. It should be recognized that patients with a right ventricular infarction often require a high central venous pressure to maintain organ perfusion.

Unstable angina

Unstable angina presents as chest pain that increases in frequency, occurs at rest, and is not associated with the enzyme and ECG changes associated with MI. It is distressing to the patient and may proceed to MI.

Immediate management

Because the other principal condition in the differential diagnosis is myocardial infarction, the initial treatment is the same: high-flow oxygen, IV access, and oral aspirin. Give sublingual GTN and if this does not relieve the pain consider IV opioids. Pain control is essential not only for humanitarian reasons but also to prevent the increased myocardial oxygen demand associated with a stress response. All these patients should be referred to the CCU for admission, but their pain should be relieved before transfer.

Further management

Nitrates Sublingual GTN should be given initially. If this does not relieve the pain or if the pain recurs start an IV infusion of GTN or isosorbide dinitrate. Nitrates cause venodilation and lower the blood pressure so the infusion rate should be titrated to relieve pain without causing hypotension. If the systolic pressure falls below 100 mmHg or the patient's urine output drops below 30 ml the infusion rate should be reduced.

Box 3.3 Management of unstable angina with nitrates

	Dilution	Starting dose	Max. dose
GTN	1 mg/ml	1 ml/h	20 ml/h
Isosorbide dinitrate	0.5 mg/ml	4 ml/h	40 ml/h

Beta blockers If there is no contraindication, such as bronchospasm, heart failure or bradycardia, a betablocker, such as atenolol 50 mg once daily, should be started.

Calcium-channel blockers If the patient continues to have pain start nifedipine 10–20 mg thrice daily. This should only be given with a beta blocker to prevent reflex tachycardia.

Heparin Start an infusion of heparin to keep the APTT twice the control value.

Coronary angiography If the pain is not controlled by quadruple therapy (nitrates, beta-blockers, calcium-channel blockers, and heparin), refer for urgent angiography.

Investigation

12-lead ECG, chest X-ray, full blood count, urea and electrolytes, and cardiac enzymes, should be performed in the A&E department.

Aortic dissection

Aortic dissection should be considered in all patients with chest pain.

Clinical features

Chest pain Typically a severe tearing pain, of sudden onset, radiating to the centre of the back. It may move site as the dissection progresses.

Signs of end organ ischaemia Acute myocardial infarction due to occlusion of origin of coronary artery. Absent peripheral pulses or a difference in limb blood pressures, hemiparesis, or impaired conscious level due to carotid artery occlusion, paraparesis due to spinal cord ischaemia, anuria from renal artery occlusion, or gut ischaemia may occur.

Physical effects of dissection Aortic regurgitation, haemopericardium, and cardiac tamponade, left haemothorax.

Investigation

Chest X-ray Many radiological signs of aortic dissection have been described. The most common are mediastinal widening (on an erect PA Chest X-ray) and a left pleural effusion.

Trans-oesophageal echocardiography, Computed tomagraphy (CT), and magnetic resonance imaging (MRI) may diagnose aortic dissection but are not 100 per cent sensitive.

Aortography is the gold standard for the diagnosis of aortic dissection. It is usually used prior to surgery even when dissection has been diagnosed by another means.

Treatment

- All patients presenting with chest pain require high-flow oxygen and IV access.
- Analgesia should be given using titrated IV opioids.
- Refer the patient to your regional cardiothoracic surgical unit.
- Treat cardiac tamponade if present.
- If the blood pressure is elevated it will need to be controlled with IV agents (after discussion with the cardiothoracic unit).

– Ensure that transfer is safe with an appropriate, senior medical escort.

Pericarditis

Causes of pericarditis

- Idiopathic or viral. There may be an associated myocarditis.
- Acute myocardial infarction—often asymptomatic. May occur 2–3 weeks post MI.
- Uraemia.
- Multisystem immune disease (e.g. SLE, rheumatoid arthritis).
- Myxoedema or malignancy (usually associated with a pericardial effusion).
- Uncommonly, bacterial, tuberculous, or fungal.

History The chest pain is typically sharp, exacerbated by breathing, worsened by lying flat, and eased by sitting forward. If a pericardial effusion develops the pain resolves but cardiac tamponade can occur causing symptoms such as breathlessness and oedema.

Examination may reveal a pericardial friction rub. This is best heard at the left sternal edge with the patient sitting forward in expiration. The rub will disappear if an effusion develops.

Investigations

ECG This typically shows concave ST elevation across the chest and limb leads. Coexisting myocarditis may cause arrhythmias.
Chest X-ray This may show an enlarged cardiac outline due to a pericardial effusion, pulmonary oedema as a result of tamponade of myocarditis or a bronchial neoplasm.
Urea and electrolytes Cardiac enzymes if MI is suspected.
Echocardiography for suspected pericardial effusion or myocarditis. It is normal in uncomplicated viral pericarditis.

Treatment

1. Assess ABC before performing a detailed history or examination. If the patient is breathless or has signs of hypo-

perfusion give high-flow oxygen and obtain IV access. If there are signs of cardiac tamponade pericardiocentesis will be required (see Chapter 16).

2. Analgesia. If the pain is severe IV opioids may be required initially. Non-steroidal anti-inflammatory drugs are very effective but remember to check for contraindications.

4. Admission will be required for severe pain, an effusion, myocarditis, or an underlying cause requiring treatment.

Assessing the patient with palpitations

All patients presenting to the A&E department with palpitations should be triaged to an area with facilities for cardiac monitoring and resuscitation.

Rapid assessment and resuscitation

1. Assess Airway, Breathing, and Circulation. The patient could have deteriorated since their arrival and may need full resuscitation.

2. If the patient's airway and breathing are satisfactory, give high-flow oxygen, secure IV access, and connect the patient to a cardiac monitor, recording a rhythm strip.

3. Assess the patient's heart rate and general condition.

It is preferable to obtain a 12-lead ECG prior to treatment in patients with arrhythmias. Exceptions to this are patients who are pulseless in whom CPR and full resuscitation is the priority, and patients who are severely compromised due to an arrhythmia who would not tolerate any delay in treatment. Once a 12-lead ECG is obtained appropriate treatment can be given, tailored to the patient's arrhythmia.

Assessment of the patient in sinus rhythm

Many patients present complaining of palpitations that have settled. Some will still have a rhythm disturbance but most will be in sinus rhythm. Connect the patient to a cardiac monitor so that if the palpitations recur in A&E the rhythm can be recorded. Ensure you understand what the patient means by palpitations, ask them to tap out the rhythm. Some patients will have simply become momentarily aware of a normal heart beat, whereas others will demonstrate the characteristic single 'dropped beat' typical of an extrasystole.

Factors associated with the palpitations that are suggestive of a serious cause include: chest pain, breathlessness, lightheadedness or faintness, loss of consciousness, previous cardiac disease or ventricular arrythmias. Any patient with these symptoms requires admission to a monitored bed on the coronary care unit or high-dependency unit. Other factors that should increase suspicion include older age group and risk factors for ischaemic heart disease. In some patients, such as those aware of a normal heart rate, palpitations are obviously benign, but if in doubt admit for observation.

Bradyarrhythmias

A bradycardia is defined as a ventricular rate less than 60/minute. Because cardiac output = stroke volume × heart rate, bradycardia can lead to a dramatic fall in cardiac output and may need immediate treatment to restore tissue perfusion.

Immediate assessment and management

The patient may have had a cardiac arrest or cerebral perfusion may be so low that the patient has lost consciousness and obstructed the airway.

1. Assess the airway and if it is obstructed open it.

2. Is the patient breathing? If not ventilate the patient using a self-inflating bag with a reservoir, connected to high-flow oxygen ensuring that the airway does not obstruct. If the patient is breathing give high-flow oxygen through a facemask.

3. Are there palpable pulses? If not, commence CPR and treat as for cardiac arrest. If there are palpable pulses insert an IV cannula and lie the patient flat.

4. Is there any easily reversible cause?

Hypoxia In children, hypoxia is the most common cause of bradycardia and although less frequent in adults it can occur and must be considered. All ill patients should be given high-flow oxygen.

Hypovolaemia Moderate hypovolaemia is associated with tachycardia but as blood loss continues the heart slows to a bradycardia which is swiftly followed by asystole and death. The treatment for hypovolaemia is volume replacement, given the massive volume loss required to cause bradycardia blood is likely to be needed. Once resuscitation is under way with oxygen and fluid the cause of the volume loss must be considered. The surgical team should be contacted urgently as surgical intervention may be required.

Vagal stimulation Vasovagal episodes may be terminated by lying the patient flat. Occasionally, pharyngeal stimulation, such as suction or passing a nasogastric or tracheal tube, causes a vagally mediated bradycardia. Stopping the procedure usually rapidly restores the heart rate to normal. Cervical dilatation causes profound vagal stimulation. The most common cause for this in A&E is a spontaneous abortion with a clot stuck in the cervix. If this is suspected speculum examination should be carried out and the clot removed under direct vision. Blood loss should be treated and the patient referred to the gynaecology team.

Sinus bradycardia

In sinus bradycardia there is a P wave preceding every QRS complex but a QRS rate below 60/minute. This may be physiological in the physically fit but in association with ischaemic heart disease or sick sinus syndrome there may be a dramatic fall in cardiac output.

Management Anyone with a ventricular rate < 40/min or symptomatic bradycardia requires treatment.

1. ABC.
2. High-flow oxygen, lie flat, IV access, and cardiac monitor.
3. Consider a precipitating cause (see above).
4. Atropine. Initially 0.5 mg IV. If ineffective this may be repeated to a total of 3 mg.
5. It is unusual for atropine to fail to increase the heart rate in sinus bradycardia. Other interventions that should be considered if this occurs include external pacing, an isoprenaline infusion and temporary transvenous pacing.

Any patient who requires treatment for bradycardia should be admitted for observation and investigation.

First-degree heart block

On the ECG every P wave is followed by a QRS complex but there is a prolonged (> 0.2 s) PR interval. It can occur as a normal variant in fit young adults. If the ventricular rate is normal no treatment is required but if it is accompanied by a bradycardia it should be managed in the same way as sinus bradycardia.

Second-degree heart block

In second-degree heart block not all the P waves are followed by QRS complexes.

Type I (Wenckebach phenomenon) In type I second-degree heart block there is progressive prolongation of the PR interval followed by a 'missed' QRS complex. Following an inferior MI this usually resolves spontaneously and if it is asymptomatic the patient can be observed on a CCU. If there are signs of hypoperfusion it should be managed as complete heart block. Following anterior MI there is a greater chance of progression to complete heart block and it is usual practice to insert a temporary pacing wire.

Type II second-degree heart block is characterized by a normal PR interval but some 'dropped' beats with a non-

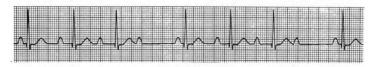

Fig 3.1 • ECG rhythm strip of type I second-degree heart block.

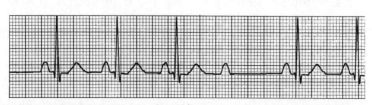

Fig 3.2 • ECG rhythm strip of type II second-degree heart block.

conducted P wave. This has a higher risk of progression to complete heart block than type I and therefore it is treated by insertion of a pacing wire. If it is accompanied by bradycardia and signs of hypoperfusion it is managed in the same way as complete heart block.

Complete or third-degree heart block

In complete heart block there is atrioventricular dissociation with no relationship between the P waves and QRS complexes.

Following acute MI
Unstable.

Because of the slow ventricular rate and the loss of co-ordinated ventricular filling patients with complete heart block may present in extremis with hypoperfusion and pulmonary oedema. Treatment must begin immediately.

1. ABC.
2. High-flow oxygen, lie flat, IV access, and cardiac monitor.
3. Atropine 0.5 mg IV repeated if necessary to a maximum of 3 mg. This is less effective than in sinus bradycardia as it may only increase the atrial rate.
4. External pacing. External pacers pass a current between two large pads on the anterior and posterior chest wall and 'capture' the ventricles at a preset rate. They are extremely effective but because they also cause contraction of the chest wall and intercostal muscles are very uncomfortable for the patient.
5. Isoprenaline infusion. Isoprenaline will increase the heart rate but at the cost of an increase in myocardial oxygen demand and a pro-arrhythmogenic effect. Because of this its

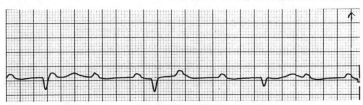

Fig 3.3 • ECG rhythm strip of complete heart block.

use is reserved for those patients who fail to respond to atropine when there is going to be a delay before pacing can be carried out. It *must* be diluted and given by infusion (e.g. 2 mg isoprenaline in 500 ml 5 per cent dextrose given at 2–10 µg/min (0.5–2.5 ml/min).

All the above are holding measures to keep the patient alive until a pacing wire can be inserted

6. Transvenous pacing. All unstable patients with complete heart block require insertion of temporary pacing wire. The technique is described in the practical procedures section (p. 216–17). It should not be carried out unsupervised by inexperienced staff. All patients who develop complete heart block following an anterior MI require a temporary pacing wire. Some patients who develop complete heart block following an inferior MI have a reasonably high ventricular rate, a normal blood pressure, and no evidence of tissue hypoperfusion. If these patients are admitted to a CCU with immediate access to pacing they can be managed without pacing.

Not following acute MI Complete heart block may present in the elderly as Stokes–Adams attacks with loss of consciousness but can also present as dizzy spells, falls, or with injuries resulting from falls. If symptomatic at presentation these patients' heart block should be treated as following acute MI, and all patients should be referred to the cardiology team for insertion of a permanent pacemaker.

Tachyarrhythmias

Broad complex tachycardia

A broad complex tachycardia (i.e. QRS width > 0.11 s) can be due either to a ventricular tachycardia or less commonly to a supraventricular tachycardia with aberrant conduction such as a bundle branch block. Before deciding which rhythm your patient has: **STOP** and assess **AIRWAY, BREATHING, AND CIRCULATION.**

Pulseless VT If a patient has a cardiac arrest (i.e. loss of

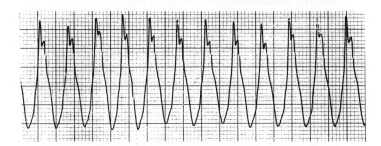

Fig 3.4 • ECG rhythm strip of broad complex tachycardia.

consciousness with no palpable pulses) with a broad complex tachycardia on the cardiac monitor this should be treated in exactly the same way as ventricular fibrillation, by a precordial thump, followed if this is unsuccessful by an unsynchronized 200 J shock. See the cardiac arrest section (p. 8–9) for the VF/Pulseless VT algorithm. The sooner the shock can be delivered the more likely it is to be successful so it should not be delayed.

Unstable broad complex tachycardia Both ventricular and supraventricular tachycardia are treated the same in an unstable patient with palpable pulses but evidence of hypoperfusion.

- Ensure a clear airway and give high-flow oxygen by face-mask.
- Establish IV access rapidly, if possible.
- Synchronized cardioversion (under sedation or general anaesthetic if the patient is conscious).
- Start with 100 J, progressing to 200 J and 360 J if this is unsuccessful.

Box 3.4 **Indications for DC cardioversion**

- Pulseless VT: (unsynchronized, follow VF algorithm)
- Hypotension: BP < 100 systolic
- Signs of pulmonary oedema
- Symptoms such as breathlessness or chest pain
- Failure of drug therapy

'Stable' broad complex tachycardia

In asymptomatic patients with no signs of hypoperfusion,

- Give high-flow oxygen.
- Establish IV access.
- Take a 12-lead ECG to distinguish between VT and SVT. Compare it with previous ECG records, if this can be done immediately.

Box 3.5 VT: diagnostic and suggestive features

Features diagnostic of VT

- Capture beats (sinus beat seen during broad complex tachycardia)
- Fusion beats—bizarre complexes from the fusion of sinus and ventricular beats
- Independent atrial activity

Features suggestive of VT
- History of ventricular arrhythmias
- History of ischaemic heart disease
- Age > 45 years
- Gross haemodynamic compromise
- ECG concordance (i.e. similar QRS axis on all chest leads)

If in doubt treat as VT

Drug treatment of VT Does the patient need any treatment? Short self-terminating asymptomatic runs of VT may be best treated by correcting electrolyte imbalance rather than giving drugs.

1. Check K⁺ and correct any abnormality.

2. Consider giving magnesium sulphate 10 ml 50 per cent over 1 hour for patients with hypokalaemia, on long-term diuretics, or with impaired LV function.

3. The drug of first choice is **lignocaine** 50 mg over 2 minutes, repeated, if necessary, every 5 minutes to a total of 200 mg. An infusion at 2 mg/min is started with the first bolus. After 3 hours the rate is slowed to 1.5 mg/min and continued for 24 hours.

If lignocaine is not successful then confer with the CCU rather than committing the patient to your favourite drug. Because all anti-arrhythmic drugs are negatively inotropic and pro-arrhythmogenic no more than 2 drugs should be given before resorting to cardioversion.

Box 3.6 Treatment that can be used for 'stable' VT

- Synchronized DC cardioversion
- Amiodarone: 900 mg over 90 min through central line
- Flecainide: 2 mg/Kg over 10–30 min
- Mexilitine: 100–250 mg over 10 min
- Procainamide: 500–600 mg over 30 min
- Overdrive pacing (through transvenous wire)

If at any stage the patient should deteriorate, developing signs of hypoperfusion treat with cardioversion.

Supraventricular tachycardia

The term 'supraventricular tachycardia' (SVT) includes any tachyarrhythmia originating above the His bundle division. It typically leads to narrow (< 0.11 s) complexes on the ECG, although in the presence of a bundle branch block or an accessory pathway, the complexes may be broad (see the section on broad complex tachycardia, p. 32.)

Initial assessment and resuscitation
- If the patient has an obstructed airway due to a reduced conscious level clear it.
- Give high-flow oxygen. If the patient is not breathing ventilate with 100 per cent oxygen.
- Check there is a palpable pulse. If not start CPR. Obtain IV access.
- Quickly assess whether this is an SVT or a sinus tachycardia due to hypovolaemia which will require volume resuscitation.
- Make a rapid assessment of whether there are signs of hypoperfusion.

Unstable SVT   If the patient has:
- hypotension;

- pulmonary oedema;
- symptoms such as chest pain or breathlessness.

DC cardioversion should be performed (with sedation if the patient is conscious). 100 J may be adequate but if this is not effective increase to 200 J and then 360 J.

Adenosine works rapidly and because of its short half-life does not cause residual myocardial suppression. If it is *immediately* available it may be used in the unstable patient. It is given by rapid IV bolus, the doses given are: 3 mg followed, if this is unsuccessful, by 6 mg, 12 mg, and 18 mg. Adenosine is contraindicated in asthma. If this is ineffective DC cardioversion should be performed.

No other drug should be used in the unstable patient with SVT.

'Stable' SVT Many patients, particularly those without underlying cardiac disease, tolerate SVT well.

Vagal manoeuvres Vagal stimulation can terminate an SVT.

Valsalva This is the most effective of the vagal manoeuvres. Put the barrel of a 10 ml syringe in the tubing of a sphygmomanometer and ask the patient to blow the mercury up to 40 mmHg and keep it there as long as they can.

Carotid sinus massage This should not be performed without careful considerations as it carries a risk of embolic stroke. It is contraindicated if there is a carotid bruit and should be avoided in those with known atheroma and in any patient over 60. It must only be performed on one side of the neck.

Diving response Immersing the face including eyes, nose, and mouth in a bowl of ice and water may terminate an SVT. It is unpleasant for the patient and co-operation is essential.

Adenosine This is given as suggested above i.e. 3 mg IV, followed, if unsuccessful, by 6 mg, 12 mg, then 18 mg IV given rapidly, followed by a saline flush. Any effect will be seen within 1 minute. It frequently causes momentary nausea or chest pain that can be very distressing so warn the patient. It may also cause momentary sinus arrest but its short half-life means these are transient. It is contraindicated in asthma.

If adenosine is ineffective consider synchronized DC cardioversion.

Synchronized DC cardioversion With sedation. 50 J initially, increased if this is unsuccessful.

If the above treatments are unsuccessful you should refer the patient to the CCU rather than committing them to your favourite drug.

Other treatments that can be used for 'stable' SVT.

Verapamil This is less effective than adenosine and can have serious adverse effects.

- Never give verapamil if there is the slightest possibility the patient has VT. It will cause massive decompensation and possibly cardiac arrest.
- Verapamil is a strongly negative inotropic. Do not administer if there is any history or signs of cardiac failure.
- Verapamil must not be used in any patient on beta blockers as they interact to cause heart block and cardiac failure.

Verapamil should only be given after discussion with the cardiology team. If you are *certain* the rhythm is narrow complexed and an SVT give 5 mg, repeated after 3 minutes if it is not successful.

Atenolol 2.5 mg repeated 3–4 times at 10-minute intervals (NOT if the patient has received verapamil).

Amiodarone 900 mg given over 1 hour through the central line.

Flecainide 2 mg/kg over 10–30 minutes.

Overdrive pacing (through transvenous wire).

Acute left ventricular failure

In acute left ventricular failure, as in any ill patient, before taking a history or doing a full examination:

- Assess ABC.
- Give high-flow oxygen.
- Establish IV access and connect to an ECG monitor.

No history may be obtainable. Examination shows a distressed patient, trying to sit upright or a comatose patient with froth in their airway. There is a tachycardia and on auscultation a gallop rhythm and crackles in the lung fields. Wheeze may be present.

What is the underlying cause?

While initiating treatment exclude an underlying cause requiring specific management:

- *Tachyarrhythmias* (i.e. VT or SVT) Treat with DC cardioversion.
- *Acute MI* Consider thrombolysis if there is no contraindication (see p. 22).
- *Anaemia*
- *Valvular dysfunction* Mitral or aortic valve dysfunction can precipitate severe pulmonary oedema. Refer to the cardiologists for consideration of surgical repair.
- *Endocarditis* Usually with valve dysfunction. Refer to the cardiology team.
- *Faulty artificial valve* If a patient with an artificial valve presents with acute pulmonary oedema it may be due to valve failure. Give oxygen and immediately refer to your local cardiothoracic surgery centre. Transfer urgently with an experienced transfer team.

Investigation

Chest X-ray confirms the clinical diagnosis of pulmonary oedema. May show a broken artificial valve.

ECG diagnoses arrhythmias and may show an acute MI.

Urea and electrolytes, glucose Hypokalaemia and hyperglycaemia are common and must be corrected.

Arterial blood gases may be required to assess the need for assisted ventilation.

Treatment

Assess and treat ABC Is the patient unconscious? have they had a cardiac arrest?

Sit the patient upright. Give high-flow oxygen, establish IV access, connect to a cardiac monitor.

Loop diuretic Frusemide 80–120 mg or bumetanide 2–3 mg IV bolus. These drugs cause venodilation, reducing preload as well as having a diuretic action. The dose required will depend on the patient's maintainance dose of diuretic.

Opioids Morphine 5–10 mg, or diamorphine 2.5–5 mg IV

with anti-emetic. Opioids reduce preload through venodilation and relieve distress. They should be diluted and given by titration against response to avoid depression of conscious level and respiratory effort

IV nitrates GTN 1–12 mg/h or isosorbide dinitrate 2–10 mg/h, given through an infusion pump. These drugs are powerful venodilators and the dose must be adjusted to avoid hypotension.

Patients with severe LVF should have a urethral catheter passed to monitor urine output and should be admitted to a high-dependency unit.

Venesection of 300 ml of blood into a standard blood collection set reduces preload and therefore may be of benefit in LVF. It is less frequently required since the development of powerful venodilators but may be life-saving in a critically ill patient where all other treatment has failed to produce any improvement.

ITU management Most patients respond to the above treatment but a small number do not. These patients need to be referred to the ITU.

Inotropes. The hypotensive patient with LVF is likely to require a combination of IV nitrates and inotropic agents, such as dobutamine or adrenaline. These should only be used with invasive monitoring of pulmonary artery wedge pressure.

Ventilation. Breathing, particularly in the presence of pulmonary oedema takes a lot of effort. Ventilation with positive end-expiratory pressure may be required.

Severe hypertension

Emergency control of hypertension in the A&E department is very rarely required and should always be done in conjunction with a senior member of the in-patient team. Most patients, even those with severe hypertension can be more safely managed with oral agents on the ward.

In severe hypertension (see Box 3.7), treatment with IV agents may be necessary.

> **Box 3.7 Severe hypertension**
>
> - Malignant hypertension with encephalopathy, acute renal failure, or LVF
> - Aortic dissection
> - Subarachnoid haemorrhage
> - Eclampsia. **Urgent** referral to the obstetric team is required if the baby and mother are to survive. Do not delay your referral.

Urgent investigation

- Check the blood pressure in both arms yourself (as you should for every patient!).
- Urea and electrolytes.
- Urinalysis and urine microscopy
- Chest X-ray—for cardiomegaly and pulmonary oedema.
- ECG—for evidence of long-standing hypertension and acute ischaemia.

Treatment

1. ABC. The patient may have an impaired conscious level. Assess and treat ABC.
2. Give high-flow oxygen and obtain secure IV access.
3. Refer to a senior member of the relevant in-patient team or ITU before using any drugs to lower the blood pressure. The blood pressure must be lowered in a controlled manner as precipitate falls, even to 'normal' levels can cause cerebral and renal ischaemia.

Agents that can be used by experience staff include:

Sodium nitroprusside is a highly effective hypotensive agent with a short half-life but it can cause severe hypotension and high doses can lead to accumulation of cyanide. It should never be used by inexperienced staff and **must** be given through a syringe driver. Starting dose 0.3–1 µg/kg/min (usual range 0.5–6 µg/kg/min).

Labetolol, diluted to 1 mg/ml. Start at 15 mg/h through a syringe pump. Double every 15–30 minutes as required. Check for contraindications (e.g. LVF).

Cardiac tamponade

Cardiac tamponade occurs when the pericardium fills with fluid or blood and prevents cardiac filling. It may result from blunt or penetrating trauma but medical causes include uraemia, malignancy, arteritis, pericarditis, myxoedema, and MI.

Diagnosis

Cardiac tamponade should be suspected in EMD cardiac arrest with dilated neck veins.

Box 3.8 Signs of cardiac tamponade

- Tachycardia with hypotension. The heart sounds are said to be muffled but this is hard to detect in a noisy A&E department
- Pulsus paradoxus (blood pressure falls on inspiration)
- High jugular venous pressure, which rises on inspiration (Kussmaul's sign) not seen with coexisting hypovolaemia.

Investigation

- Chest X-ray may show an enlarged globular cardiac shadow.
- ECG shows small complexes.
- Echocardiography or ultrasound clearly demonstrates the pericardial effusion.

Treatment

1. Assess and treat ABC. Has the patient had a cardiac arrest? If so commence CPR and perform immediate pericardiocentesis.
2. High-flow oxygen, establish IV access, connect to a cardiac monitor.
3 Pericardiocentesis is the definitive treatment of cardiac tamponade. Ideally this is done under ultrasound control to minimize the risk of myocardial injury but in a patient in extremis

it can be done with ECG monitoring alone. Even removal of a small amount of fluid can restore an effective circulation.

4 If the cardiac is traumatic in origin, urgent thoracotomy is required but pericardiocentesis may remove enough blood to keep the patient alive until a surgeon arrives. (The technique of pericardiocentesis is described in detail on p. 217–19.)

Further reading

1. Handley, A. J. and Swain, A. (ed.) (1994). *Advanced life support Manual.*
2. McMurray, J. and Rankin, A. (1994). Treatment of myocardial infarction, unstable angina and angina pectoris. *British Medical Journal*, **309**, 1434–50
3. Weston, C. F., Penny, W. J., *et al.* (1994). *Guidelines for the early management of patients with acute myocardial infarction. British Medical Journal*, **308**, 767–71.
4. Brugada, P., Gursoy, S. *et al.* (1993). Investigation of palpitations. *Lancet*, **341**, 1254–8.
5. Murgatroyd, F. D. and Camm, A. J. (1993). Atrial arrhythmias. *Lancet*, **341**, 1317–22.
6. Shenasa, M., Borggrefe, M., *et al.* (1993). Ventricular tachycardia. *Lancet*, **341**, 1512–19.
7. McMurray, J. and Rankin, A. (1994). Treatment of heart failure and atrial fibrillation and arrhythmias. *British Medical Journal*, **309**, 1631–5.

CHAPTER 4

Respiratory emergencies

The breathless patient

Acute dyspnoea is a distressing symptom for patients. The feeling of breathlessness arises when there is an imbalance between the work of breathing and the physiological result (i.e. the arterial partial pressures of oxygen and carbon dioxide). The central chemoreceptors are acutely sensitive to any rise in the arterial partial pressure of CO_2 (PCO_2), but require the arterial partial pressure of oxygen (PO_2) to fall to 8 kPa (60 mmHg), or less, before there are any major effects. Thus, after exercise, a patient may have a very high minute volume, but no sensation of dyspnoea, but an asthmatic may have a normal minute volume and relatively normal blood gases with a marked sensation of dyspnoea because of the work involved. The haemoglobin–oxygen dissociation curve is S-shaped and an arterial oxygen saturation of 90 per cent corresponds to a PO_2 of 8 kPa. Below this figure, there is a rapid fall in PO_2 as saturations decrease.

Arterial blood gas measurements (ABG) are invasive, painful, and may take some time to obtain. Pulse oximetry allows much easier monitoring of oxygenation in an emergency but it is important to realize that it gives no indication of PCO_2. A patient may have normal oxygen saturation with severe hypercapnia and respiratory acidosis. It is therefore essential that all critically ill patients have arterial blood gas analysis performed.

Assessment

All breathless patients must be triaged to an appropriate area and assessed rapidly. Those with marked distress must be seen immediately in the resuscitation room using a team approach if possible. Vital signs, pulse oximetry, and peak expiratory flow rate (for asthmatics) should be recorded on arrival.

Airway Although much less common in adults than children, upper airway obstruction may be rapidly fatal. Total airway obstruction in characterized by absence of airflow despite obvious respiratory effort and, in the conscious patient, usually bilateral grasping of the throat. The hallmark of incomplete obstruction is stridor, the noise of restricted airflow in the upper airways. Stridor is loudest on inspiration, when negative airway pressure causes partial collapse of the extrathoracic airway. This distinguishes it from wheeze which originates in the small airways, and is loudest on expiration when raised intrathoracic pressure collapses the intrapulmonary bronchi.

Breathing Clinically, breathing is assessed in terms of rate, depth, effort, and pulmonary findings. The range of normal respiratory rate (10–24 per minute) in wide, but any adult patient outside this range is likely to have significant pathology. Depth is usually assessed subjectively as is respiratory effort, based on tracheal and intercostal space in-drawing, use of accessory muscles, and patient distress.

A simple measure of dyspnoea is the ability to speak: whether the patient can say sentences, 2–3 words, or only single words between breaths.

Auscultation of the chest should assist in localizing the pathology except when there is little airflow; beware the 'silent chest'. Pay attention to the loudness of the breath sounds (sometimes incorrectly called air entry). The difference between sides in pneumothorax or pleural effusion may be subtle.

A bilaterally quiet chest means very poor airflow, and

wheezes and crackles may actually get louder as treatment increases flow. Localized crackles and abnormal breath sounds suggest consolidation, but bibasal 'moist' crackles do not always indicate pulmonary oedema—bilateral infection and/or fibrotic changes may be the cause.

Cyanosis is an unreliable sign in respiratory assessment. It is said to require at least 5 g/100 ml of desaturated haemoglobin, and may require more in the setting of a poorly lit A&E department, hence it cannot occur in severely anaemic patients. Cyanosis indicates respiratory or cardiovascular pathology, which may be acute or chronic. The absence of cyanosis is not indicative of health. Pulse oximetry is more effective at assessing hypoxia.

Spirometry, or peak expiratory flow (PEF) measurement, is mandatory whenever any limitation to pulmonary function is suspected. It is particularly important to attempt to obtain PEF (or spirometry) prior to starting nebulized therapy in obstructive airways disease. Be aware that not all respiratory presentations involve lung pathology. In addition to cardiac causes, metabolic acidosis (especially diabetic ketoacidosis, see Chapter 7) can cause profound dyspnoea.

Circulation Cardiac and pulmonary function are intimately linked, both anatomically and physiologically. Disease in either system may present with signs and symptoms in the other, so cardiovascular assessment is essential in the breathless patient. Patients with severe left ventricular failure may wheeze and other cardiovascular diseases may present with predominantly respiratory signs. The patient's history normally indicates if congenital heart disease is the cause of cyanosis, but pulmonary oedema due to left ventricular dysfunction can be difficult to distinguish from infection, and they may coexist.

Investigation

Pulse oximetry and PEF (or spirometry) are bedside tests appropriate in all patients with respiratory emergencies. Be careful in interpreting the oxygen saturation in a patient on oxygen therapy, since it is possible to have severe hypoventilation and a high oxygen saturation if supplemental oxygen is being administered. Similarly, be careful to interpret PEF (or

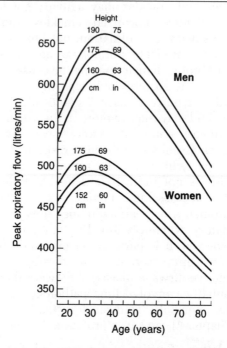

Fig. 4.1 • Peak expiratory flow in normal adults. From British Thoracic Society Guidelines (1993). *British Medical Journal*, **306**, 779.

spirometry) in the light of the patient's age, size, and past history, comparing the results to previous figures if available (see Fig. 4.1).

A chest X-ray is mandatory. When reviewing this, carefully scan each lung field, look specifically for small pneumothoraces (best seen at the apex if it is an upright film) follow the cardiac and diaphragmatic borders looking for blurring, and remember to look behind the heart on the lateral film.

Despite the development of pulse oximetry, arterial blood gases remain a useful assessment of respiratory function, and should be performed on all ill respiratory patients. The P_{CO_2} is a measure of minute volume. It may be chronically raised in those with long-term, severe respiratory disease, when it is associated with a markedly elevated calculated bicarbonate level. The P_{O_2} is a measure of alveolar to capillary oxygen diffusion, and of intrapulmonary or extrapulmonary shunt ('ventilation: perfusion mismatch'). It is essential to know the inspired oxygen concentration when interpreting the Po_2. In an ill patient take blood for blood gases on a fixed concentration oxygen delivery system, do not risk hypoxia by letting the patient breathe air.

Respiratory failure is defined as $Po_2 < 8kPa$ (< 60 mmHg) and is subdivided into Type I, due to inadequate gas exchange, with normal or low Pco_2 (< 5.9 kPa or < 45 mmHg), and Type II, due to hypoventilation, with a raised Pco_2.

Other investigations are normally undertaken on patients who are sufficiently ill to be admitted. Full blood examination, urea, creatinine, electrolytes, blood cultures if febrile, and sputum examination are generally considered routine.

Treatment

Specific therapy obviously depends on the particular condition but there are some important principles as listed below.

Sit the patient up Bracing the arms helps the use of accessory muscles and improves ventilation. Any dyspnoeic patient who is capable of sitting should be nursed in that position.

Give oxygen All patients who are hypoxic benefit from supplemental oxygen. Patients with chronic hypercarbia may have their respiratory drive reduced if hypoxia is corrected, but this is an indication for close monitoring and measurement of arterial blood gases, not a reason to withhold oxygen from an hypoxic patient.

Oxygen delivery systems are classified as fixed performance (Venturi systems, with or without mixing tube) and variable performance (nasal prongs, facemasks). High-flow (8–12/min) through a disposable facemask delivers a maximum of 40–50 per cent oxygen, and is the usual starting point for oxygen therapy. Disposable reservoir masks deliver a maximum of 80 per cent oxygen, and only anaesthetic circuits achieve near 100 per cent oxygen. Note that patients with a high minute volume may find low-flow oxygen delivery systems distressing if air inflow is restricted.

Monitoring Ill patients should be monitored by pulse oximetry and regular observation of respiratory rate, pulse rate, blood pressure, and conscious state. As noted above, normal oxygen saturation on oxygen therapy does not exclude hypercarbia, and ABG measurements are indicated in patients with an abnormal conscious level or chronic airways disease.

Intensive care   Patients who do not respond appropriately to adequate therapy, or are becoming exhausted due to prolonged respiratory effort, need to be considered for intubation and ventilation in the ITU. Seek advice of your ITU early, but recognize that patients whose lungs have been destroyed by a chronic disease process will not be regarded as candidates for intubation and ventilation since there may be no prospect of recovery to an adequate quality of life (see Chapter 14).

Upper airway obstruction

The adult airway is much larger than the child's, so that it requires greater external compressive force or a larger internal obstruction to cause a problem. Hence, airway obstruction is a less common presentation than in paediatrics (see *Accidents and emergencies in children*, in this series), although the pathology is similar. Inhaled foreign bodies are a common cause, but pharyngeal swelling due to infection, trauma (particularly in patients with coagulopathy or taking warfarin), or malignancy may be responsible.

Treatment

An inhaled foreign body, usually poorly chewed food, may cause complete or partial obstruction. Complete obstruction is an emergency which occurs in the community. If no large foreign body can be seen in the mouth and removed, complete airway obstruction must be treated urgently by raising intrathoracic pressure in order to expel the obstruction. Use back blows or the Heimlich manoeuvre. If this fails, an emergency surgical airway is indicated (see Chapter 16). In patients presenting with partial airway obstruction it is important to attempt to distinguish the level of obstruction. Patients localize relatively poorly in the upper airway, and very poorly in the lower airway and oesophagus.

A small foreign body that caused some upper airway problem when inhaled may have passed into the lungs by the time of presentation. Such a lesion is said to cause a monophonic wheeze, but this is an uncommon finding. It may be localized by X-ray (showing the foreign body if radioopaque, or segmental collapse of the lung), and requires admission for rigid bronchoscopy for removal.

Larger and often sharper foreign bodies may lodge anywhere in the posterior pharynx or upper larynx. Removal may be carried out by senior staff in the A&E department by direct laryngoscopy under local anaesthetic. If the patient can tolerate it, position them and use local anaesthetic spray to attempt to visualize the vocal cords. The foreign body is best removed with Magill forceps. This procedure should not be undertaken lightly, if the foreign body is dislodged to cause complete airway obstruction, a surgical airway will be required. Some foreign bodies can only be removed under general anaesthesia. This must be performed in an operating theatre by senior anaesthetists.

If you suspect partial airway obstruction, immediately call for senior A&E, anaesthetic, and ear, nose, and throat (ENT) support. The patient will usually find the posture associated with least obstruction, so do not force them to lie down. Do not under any circumstances sedate the patient or be tempted to use anaesthetic agents yourself as this may precipitate total obstruction. Airway management in these patients is difficult, usually requiring gaseous induction and great anaesthetic expertise.

Epiglottitis

Epiglottitis is uncommon in adults, and, for the reasons outlined above, the need for airway control is much less than that in children. It is usually caused by infection, most frequently due to *Haemophilus influenzae*, but sometimes occurs in inhalational injuries and from swallowed caustic chemicals. Rarely, inflammation or trauma of another part of the pharynx, larynx, or trachea may present with the same picture. Patients present with stridor, a hoarse voice and difficulty swallowing. They may lean forward and often drool, and always complain of a severe sore throat. If you suspect this diagnosis, enlist expert help immediately. **Do not use intravenous muscle relaxants**. When transporting such a patient anywhere, including X-ray, they must be accompanied by intubation equipment and a doctor, trained in anaesthesia, able to perform tracheal intubation and cricothyroidotomy in the event of a complete obstruction.

Do not attempt to visualize the throat, as this may precipitate airway obstruction. The diagnosis should be clinical, not radiological.

If the patient does not have significant airway obstruction, a lateral soft tissue X-ray of the neck may show a swollen epiglottis, displaced posteriorly, but the X-ray will not affect management.

Treatment

Give supplemental oxygen. If the patient is hypoxic or hypoventilating because of their airway problem, they will require intubation urgently. Get senior help **immediately**.

The circulation is not normally compromised, although the patient may be a little dehydrated. Insert an IV line, and take blood for full blood count and blood cultures.

Begin broad spectrum antibiotics. Epiglottitis is usually considered an indication for third-generation cephalosporins (cefotaxime 1000 mg 6 hourly).

Patients with any significant airway obstruction need to be urgently assessed by an anaesthetist and an ear, nose, and throat (ENT) surgeon, with a view to intubation under anaesthetic.

Disposition

All patients with epiglottitis or progressive upper airway obstruction need close observation, usually in the ITU if there is any possibility of progressing to complete airway obstruction. Some will go directly from A&E to the operating theatre for intubation.

Asthma

Asthma is an inflammatory disease, in which air flow is limited by constriction of smooth muscle walls, mucosal oedema, and increased exudate into the lumen. Exacerbations of asthma may be precipitated by many factors including, infection, emotion, exercise, allergy, exposure to cold, or drugs, such as non-steroidal anti-inflammatory drugs and beta blockers. All ill patients with asthma should be seen by a doctor immediately on their arrival in the A&E department and should be treated in the resuscitation room. High-flow oxygen and nebulized salbutamol, with oxygen as the driving gas, should be started immediately, while the patient is being assessed and being connected to an ECG monitor and pulse oximeter. It is important to be aware that patients often underestimate the severity of an asthma attack and may not be distressed despite a life-threatening asthma attack. The peak expiratory flow (PEF) should be measured in all asthmatic patients, to give an objective measure of the severity of their bronchospasm.

Acute severe asthma

Asthma can kill. If your patient has any of the features of acute severe asthma call for senior help immediately.

Box 4.1 **Features of acute severe asthma**

- Cannot complete sentences in one breath
- Respiratory rate > 25 breaths/minute
- Pulse rate ≥ 110/minute
- PEF < 50% of predicted or best

Box 4.2 **Life-threatening features of severe asthma**

- PEF < 33% of predicted or best
- Silent chest, cyanosis, or feeble respiratory effort
- Bradycardia or hypotension
- Exhaustion, confusion, or coma

If the patient has ANY of the features shown in Box 4.1 and 4.2, treat as acute severe asthma and measure the arterial blood gases. (Box 4.3). No other investigations are needed for immediate management.

Box 4.3 **ABG markers of a very severe, life-threatening attack of asthma**

- Normal (5–6 kPa, 36–45 mmHg) or high PCO_2
- Severe hypoxia: $PO_2 < 8$ kPa (60 mmHg)
- Acidosis (low pH or high H^+)

Immediate management
- High-flow oxygen, above 40 per cent in all cases. In, asthma high concentration oxygen does not cause CO_2 retention.
- Salbutamol 5 mg or terbutaline 10 mg via an oxygen-driven nebulizer. This can be repeated and given continuously if the patient is very ill.
- Steroids: prednisolone 30–60 mg orally or hydrocortisone 200 mg IV.
- Do not give sedation of any kind.
- Portable chest X-ray in the resuscitation room, to exclude pneumothorax.

If there are life-threatening features:

- Add ipatropium 0.5 mg to the nebulized salbutamol/terbutaline.
- Give IV bronchodilator. Aminophylline 250 mg over 20 minutes can be given only if the patient does not take oral theophyllines. Alternatively, salbutamol or terbutaline 250 µg can be given over 10 minutes.

Subsequent management
- If the patient is improving, continue high-flow oxygen, nebulized beta agonist at least 4 hourly and refer to the in-patient medical or respiratory team for admission to a high-dependency unit.
- If the patient is not improving after 15–30 minutes continue high-flow oxygen, give nebulized beta agonist more frequently (every 15–30 minutes), add ipatropium 0.5 mg to the nebulizer and repeat 6 hourly.

• If the patient is still not improving continue oxygen and nebulized beta agonist and start an IV infusion of aminophylline (small patient (< 50 kg) 750 mg/24 h, large patient (> 80 kg) 1500 mg/24 h), or alternatively, an IV infusion of salbutamol or terbutaline.

Monitoring

1. PEF—Repeat PEF measurement 15–30 minutes after starting treatment.
2. Pulse oximetry—maintain So_2 > 92 per cent.
3. Arterial blood gases—should be repeated within 2 hours of starting treatment if the initial Po_2 < 8 kPa, unless subsequent So_2 > 92 per cent, the initial Pco_2 was normal or raised, or the patient deteriorates.

Indications for transfer to the ITU

• Deterioration in PEF, persisting or worsening hypoxia, or hypercapnia.
• Exhaustion, feeble respirations, confusion, or drowsiness.
• Coma or respiratory arrest.

During the transfer the patient should be accompanied by a doctor, prepared to intubate.

Moderate and mild asthma

Moderate asthma is defined as PEF > 50–75 per cent of predicted, with none of the features of severe or life-threatening asthma (i.e. able to talk in sentences, pulse rate < 110, respiratory rate < 25). *Mild asthma* is defined as PEF > 75 per cent of predicted, with none of the features of severe asthma. During the treatment of mild or moderate asthma continually reassess the patient, looking for any deterioration. If the patient develops any of the features of severe or life-threatening asthma start treatment as outlined above and call for help from senior staff.

Chronic obstructive airways disease (COAD or COPD)

Chronic obstructive airways disease describes the syndrome of emphysema and chronic bronchitis. Most commonly as a

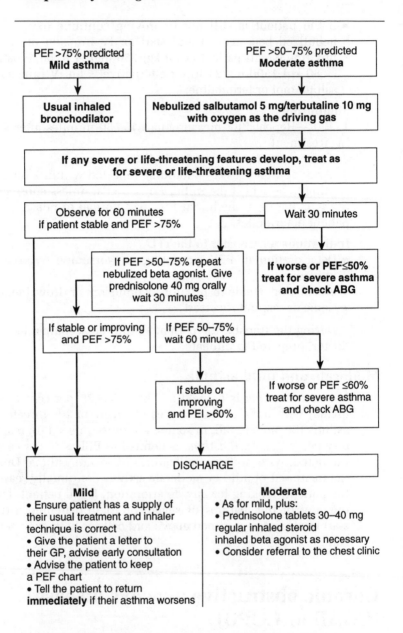

Fig. 4.2 • The management of asthma. From British Thoracic Society Guidelines (1993). *British Medical Journal*, **306**, 782.

result of cigarette use, these patients have lost both lung tissue and pulmonary compliance. In addition, some have reactive airways disease. Classically these patients are divided into 'pink puffers' and 'blue bloaters' but their emergency management is little different. Bear in mind, however, that blue bloaters with chronic bronchitis, right ventricular failure, and hypercarpia tend to have multiple exacerbations and slowly worsen, whereas pink puffers with emphysema tend to have a small number of very severe exacerbations before death.

The usual presentation of 'exacerbation of COAD' is usually a viral-mediated bronchitis/pneumonitis, with secondary airway mucosal oedema, increased mucosal exudate, and bronchospasm. Diseased lungs are more prone to such infections, and lack of pulmonary reserve makes a small decrease in function potentially fatal. However, diseased lungs are also more prone to bacterial bronchitis, segmental collapse, pneumonia and pneumothorax, and the association of smoking and heart disease makes these patients more prone to left ventricular failure. Thus, each of these pathologies must be specifically excluded before a patient is categorized simply as an 'exacerbation of COAD'.

The Treatment is for infection plus asthma. The ill patient should be assessed and managed in the same way as an ill asthmatic (see p. 51–3).

Immediate assessment and management

1. Airway, Breathing, Circulation.
2. Give supplemental oxygen, starting at 24 per cent. Obtain PEF (or spirometry) if possible. Sit the patient up, obtain IV access. Use pulse oximetry if available but remember this does not detect hypercarbia.
3. Begin treatment of bronchospasm with nebulized salbutamol 5 mg or terbutaline 10 mg. Consider adding ipratropium 0.5 mg.
4. Obtain a chest X-ray and arterial blood gases, preferably on a known inspired concentration of oxygen. Take blood for full blood count, urea and electrolytes, blood cultures if febrile, serum theophylline, if indicated, and perform an ECG.
6. Treat any precipitating factor, such as pneumothorax and ischaemic heart disease.
7. Begin antibiotic therapy. If is appropriate to follow local

prescribing patterns, but a reasonable starting point is amoxycillin 500 mg 8 hourly.

8. Start steroids—prednisolone 30 mg, or 300 mg IV hydrocortisone if the patient cannot take oral drugs.

Disposition

Virtually all patients who present to A&E departments with acute exacerbations require admission. All patients with COAD should have regular contact with their GP, and start oral antibiotic therapy and steroids at the earliest sign of an exacerbation. The majority do so, and, like asthmatics, know their disease well, presenting to hospital only when admission is needed.

A few patients may use the A&E department for ongoing care which should be discouraged: irregular contact with different doctors is not in the patient's interest.

Patients who present early with previously untreated exacerbations may be suitable for out-patient management if they appear well and have PEF > 75 per cent of their normal. Use oral antibiotics (amoxycillin 500 mg 8 hourly) and steroids— 30 mg prednisolone daily for 7 days. Ensure that GP follow-up is arranged and that the patient is willing and able to return if they worsen.

Cautionary points

Loss of respiratory drive and resultant hypercarbia is an uncommon but real problem in patients with COAD who are given oxygen. Ideally, patients with chronic hypercarbia should be managed on pulse oximetry with the inspired oxygen varied to maintain So_2 90–92 per cent. This is a difficult task, and the range 88–94 per cent is reasonable.

Seek advice early from an intensivist if the patient is deteriorating. It is undesirable to intubate a patient on the point of death and then discover that they are not suitable for intensive care.

Pneumonia

A wide variety of organisms cause a wider variety of signs and symptoms in pneumonia. The clinical examination, X-ray changes, and arterial blood gases may not be predictably related, particularly if the patient has pre-existing lung disease or the organism is 'atypical'. *Streptococcus pneumoniae* (pneumococcus) is the usual cause of lobar pneumonia, and also among the most common causes of bronchopneumonia, although the difference between these two classical scenarios is unimportant in the emergency setting. The questions requiring a answer are whether the patient needs admission, and which antibiotic is appropriate. The presentation may range from 'classical' with sepsis, fever, pleuritic chest pain, and cough with purulent sputum, through to simply confusion or fever, when pneumonia is identified on examination or investigation (see also Chapters 6 and 12). These presentations are most likely in the elderly (Chapter 13) and the immuno-compromised (Chapter 11).

'Typical' pneumonia occurs either primarily in healthy individuals (most commonly due to pneumococcus), or secondarily in those with pre-existing lung disease, severe viral infection (especially influenza), or aspiration. 'Atypical' pneumonia is usually community-acquired, and the chest X-ray looks much worse than the patient's symptoms would suggest. In most cases, the patients are well enough to wait for some confirmatory tests (e.g. Gram stain sputum, cold agglutinins) prior to antibiotic choice, or to be treated 'blindly' orally, and discharged.

It is more difficult to choose an antibiotic for 'blind' treatment in the severely ill or immuno-compromised patient, when treatment is mandatory as soon as cultures are sent. Table 4.1 offers guidance only, and any decisions should be made in consultation with your hospital's specialists according to local organisms and local practice.

Immediate assessment and management

1. Airway, Breathing, Circulation.
2. Give high-flow oxygen by facemask, obtain IV access, and monitor pulse oximetry
3. Thorough history and examination.
4. Obtain chest X-ray and arterial blood gases.
5. Full blood count, urea and electrolytes, blood cultures if febrile, 'atypical' serology (baseline sample).
6. Begin appropriate antibiotic therapy.

Table 4.1 • Antibiotics for the treatment of pneumonia

Clinical picture	Likely organisms	Antibiotic
Previously healthy Primary pneumonia	Pneumococcus	Amoxycillin 500 mg IV 8 hourly or 500 mg orally 8 hourly
Atypical pneumonia	Chlamydia Mycoplasma Legionella	Erythromycin [b] 250 mg orally 6 hourly or 500 mg IV 12 hourly
Pre-existing COAD or bronchiectasis	Haemophillus pneumococcus	Amoxycillin IV as above
Aspiration	Pneumococcus Klebsiella/Gram-negative anaerobes	Amoxycillin[c] IV as above plus gentamicin IV 3–5 mg/kg daily plus metronidazole IV 500 mg 8 hourly
HIV[a]	Pneumocystis (see p. 161–3)	Cotrimoxazole IV or oral in 4 daily doses = trimethoprin 15mg/kg/day and sulphamethoxazole 75 mg/kg/day
Severe sepsis with pneumonia,[a]	Any organism but no time to identify	Cefotaxime IV 1000 mg 8 hourly plus erythromycin[b] 500 mg IV 6 hourly plus Gentamicin 3–5 mg/kg/day

[a] Always seek advice from appropriate specialists in your hospital if possible (but do not delay treatment).
[b] Intravenous erythromycin is painful and sclerosant and should be avoided if patient can take oral erythromycin, although this causes nausea.
[c] Co-amoxiclav is often substituted for amoxycillin but has more gastrointestinal effects.

Disposition

Indications for admission are:

• Sepsis or confusion.
• Hypoxia, $P_{O_2} < 10$ kPa (70 mmHg) on room air.
• Failure of oral therapy or contraindication (e.g. vomiting).
• Significant pre-existing lung disease (e.g. asthma, COAD, cystic fibrosis).
• Immunocompromise (e.g. leukaemia or AIDS).
• Poor home circumstances.

Pneumothorax

There has been a recent move towards much more conservative management of spontaneous pneumothorax. Refer to the

British Thoracic Society guidelines in Further reading for more details.

A simple pneumothorax occurs when a small tear in the lung surface allows air into the pleural space and then seals over. This may occur spontaneously (classically in Marfan's syndrome but commonly in tall, thin people), or as a result of asthma, emphysema, or scuba diving. Patients without underlying respiratory disease do not suffer respiratory embarrassment with a simple pneumothorax, although pleuritic pain may be severe. A tension pneumothorax occurs when a lung tear forms a flap valve, letting air into the pleural space but not out. This is life-threatening since raised intrathoracic pressure prevents venous return. However, the classic signs of hypotension, distended neck veins, and a silent chest are late and pre-terminal. Many patients with tension pneumothorax are tachypnoenic, tachycardic, and hypertensive due to sympathetic outflow. Tension pneumothorax is readily converted to open pneumothorax by use of a needle or IV catheter into the pleural space. This is a life-saving procedure.

Traumatic pneumothorax due to blunt trauma, wounds, or as a complication of central line insertion is a very different condition. It is normally associated with a continuing leak and should be managed with an intercostal drain rather than conservatively.

Never give a patient with a pneumothorax nitrous oxide (entonox) for analgesia as it will diffuse into the pleural space, and may cause a tension pneumothorax.

Tension pneumothorax

This occurs in the setting of a very ill patient, and forms part of the B in the ABC of resuscitation. It may be necessary to treat on suspicion of a tension pneumothorax, although the treatment will induce a pneumothorax if one is not present, and mandates later insertion of an intercostal catheter.

In a patient with cardiovascular and/or respiratory collapse (hypotension, cyanosis, or low oxygen saturation) and any sign of pneumothorax (dilated neck veins, tracheal deviation, reduced breath sounds, with hyperresonance) identify the side of pathology clinically without waiting for the chest X-ray.

1. Insert a 12–14 gauge intravenous catheter, long enough to reach the pleural space, directly through the 2nd intercostal

space in the mid clavicular line. Escaping air from the pleural space will be heard if the diagnosis of tension pneumothorax is correct.

2. Insert an intercostal catheter through the 4th or 5th intercostal space in the mid axillary line and connect to an underwater seal (see Chapter 16).

3. Admit the patient to hospital.

Management of simple pneumothorax

1. Airway, Breathing, Circulation.

2. Obtain chest X-ray, including expiratory film. Pneumothoraces are classified on an erect chest X-ray as:

– *Small* a rim of air around the lung occupying less than half the space to the heart border. This corresponds to a volume loss of up to 80 per cent.
– *Moderate* lung collapse halfway to the heart border.
– *Complete* no air-filled lung remaining (i.e. separate from diaphragm).

(On a supine film the diagnosis of pneumothorax is more difficult.)

3. Seek any underlying cause—asthma, COAD, scuba diving, trauma.

4. Perform drainage procedure if indicated.

Indication for intercostal catheter (chest drain) insertion (Chapter 16)

- Tension pneumothorax (see above), after emergency intercostal cannula.
- Traumatic pneumothorax.
- Failure of simple aspiration (failure to produce a small pneumothorax from a moderate or complete one).

Indications for simple aspiration

- Significant dyspnoea (i.e. deterioration in usual exercise tolerance).
- Moderate or complete collapse in patient with chronic lung disease. This includes cystic, fibrotic, bullous, or emphysematous disease, but not mild or moderate asthma.
- Complete collapse in a previously healthy individual.

In patients with bullous lung disease and emphysema it can be hard to differentiate a pneumothorax from a bulla. If you are uncertain, do not insert a chest drain or aspirate, but seek expert advice.

5. After pleural aspiration, repeat the chest X-ray in inspiration. If the moderate or complete pneumothorax has not been converted into a small pneumothorax, insert a chest drain.

6. Treat any underlying cause, especially asthma or COAD.

Indications for admission

1. Chest drain inserted.

2. Chronic lung disease, whether or not aspirated. Overnight observation may be all that is required.

3. Underlying cause requiring admission.

On discharge, ensure the patient knows to return if they deteriorate; will not fly or scuba dive; and has a letter for their general practitioner.

Pulmonary embolus

Pulmonary embolus is a difficult condition to diagnose, and a high degree of suspicion is required. The clinical signs of deep venous thrombosis are notoriously unreliable.

Box 4.4 **Risk factors for thrombo-embolism**

- Recent surgery or illness
- Prolonged bus or airline trips
- Smoking
- Oral contraceptive or anabolic steroid use

The findings described as classical in many textbooks are relatively late markers of pulmonary embolus. The history is normally of a sudden onset of shortness of breath, possibly with pleuritic chest pain. Examination may reveal nothing, sinus tachycardia, or, if there has been a large embolus or multiple emboli, signs of right ventricular strain (raised jugular venous pressure, loud P2). If pulmonary infarcts occur a pleural rub and/or chest X-ray findings may be of assistance. The ECG commonly shows a sinus tachycardia only, but in larger emboli right ventricular strain (S1, Q3, T3, and T wave changes in V1–3) pattern is seen, sometimes RBBB or R-axis deviation. Arterial blood gases are a sensitive but non-specific measure, results depending on the size of the embolus. Patients normally have mild hypoxia and mild hypocarbia secondary to increased minute volume.

Management of massive pulmonary embolus

This presents as cardiac arrest or sudden collapse. Follow normal cardiac arrest protocol (see Chapter 2). Cardiopulmonary resuscitation (CPR) may break up the embolus and distribute it to the periphery of the lungs with significant improvement in the patient's condition. If the patient responds to CPR, then they should be treated as for large pulmonary embolis (below) and admitted to the ITU.

Management of large pulmonary embolus

This presents as collapse, sudden respiratory distress, cyanosis and shock, with signs of acute right heart failure (raised JVP, loud P2, loud S3, R-sided ECG changes). It is a difficult condition to manage, and senior help should be sought. Unless another embolism occurs, the patient will be most ill at the start and the aim is to support them until lysis of the embolus occurs.

Airway Usually patent. Check and open by whatever means necessary.

Breathing Give high-flow oxygen preferably with reservoir bag. Use bag-valve-facemask if the patient is apnoeic. Intubation is indicated if the patient cannot maintain adequate oxygenation. Seek ITU advice early.

Circulation Obtain IV access and, if the patient is in shock and you are certain that the diagnosis is not left ventricular failure, give a fluid bolus, starting with 250 ml haemaccel. Increasing right ventricular filling pressure will help right heart output to a certain extent. Once the patient is maintaining a reasonable cardiac output, they should not receive excess fluids. If there is no response to a fluid bolus, consider inotropes in conjunction with the ITU.

Emergency measures A variety of immediate interventions have been described, including pulmonary thrombectomy and thrombolysis. Seek expert advice early. Urgent venous ligation or passage of an inferior vena caval 'umbrella' may prevent further emboli and should also be considered.

5. *Ongoing management* As for small pulmonary embolism, below. Generally admit to the ITU.

Management of small pulmonary embolus

As noted above, a high index of suspicion is required. Diagnostic tests such as ventilation/perfusion radioisotope scanning, are not available in all centres and in such situations it may be necessary to treat on clinical suspicion and to confirm the diagnosis later.

1. Airway, Breathing, Circulation.
2. Sit the patient up, give high-flow oxygen by facemask, insert IV line.
3. Take blood for fbc, U&Es, clotting profile, and arterial blood gases.
4. Obtain chest X-ray (often normal, excludes other causes) and ECG.
5. Begin anticoagulant therapy. Heparinize the patient with 5000 units heparin IV, then start an IV infusion of heparin, the dose titrated against the clotting times.
6. Give analgesia if required.

Disposition

Patients with large pulmonary emboli should be admitted to the ITU. Patients with small emboli, suspected or proven, must be admitted to hospital for both investigation (usually radioisotope ventilation/perfusion scanning) the treatment. Check your own hospital's policy if the patient had recent surgery, since some hospitals re-admit under the surgical team as a complication, and others consider this to be a new medical condition.

Massive pleural effusion

There are four major causes of pleural effusion (1) infection; (2) malignancy; (3) cardiac failure; and (4) hypoalbuminaemia. However, massive pleural effusion to the point of respiratory compromise is common only in malignancy. This may be primary or secondary, but it is most frequently seen in broncho-genic and pleural tumours. Treatment of a malignant pleural effusion is a palliative measure. For non-malignant effusions, treat the underlying cause first, with pleural aspiration only if the effusion is symptomatic.

The usual problem in the emergency setting is of a patient with known disease whose symptoms have markedly wors-ened. Thus, the history is of gradual increasing shortening of breath and the examination reveals unilateral stony dullness on percussion, with diminished or absent breath sounds. The chest X-ray makes the diagnosis clear.

Management

1. Airway, Breathing, Circulation.
2. Sit the patient up, give supplemental oxygen and analgesia.
3. Urgent drainage of massive pleural effusion should only be undertaken if the patient has severe respiratory symptoms. It depletes the patient of protein, causes potentially dangerous mediastinal shifting, and is commonly only a palliative procedure.

With the patient connected to a cardiac monitor, using a 12–14 gauge intravenous catheter, 3-way tap and a 50 ml syringe, tap the effusion through the 8th intercostal space posteriorly. It is advisable to withdraw no more than 500–1000 ml at any one time (maximum of 2 litres/24 h). Chest discomfort is caused by shifts in the mediastinum with resultant traction on the great vessels, a potentially serious complication, if it develops stop aspiration. It may be decided to insert a chest drain. If the cause is not known sent the fluid for glucose, protein, cytology, Gram stain, C&S, acid-fast bacilli, and possibly immunology. Pleural biopsy may need to be done later but this is not an emergency procedure.
4. Repeat the chest X-ray to assess resolution, complications such as pneumothorax, and allow examination of the lung fields, and admit the patient under the in-patient team.

Hyperventilation

The hyperventilation syndrome is a distressing combination of subjective dyspnoea, increased minute volume, hypocarbia, respiratory alkalosis, and secondary symptoms due to a decrease in serum ionized calcium. It is of psychogenic origin and may affect patients of any age or either sex.

The symptoms of hyperventilation syndrome consist of fear, shortness of breath, light-headedness, perioral, and digital paraesthesia, and in extreme cases, tetany. These symptoms cause the patient to further increase respiration, leading to a vicious circle. The signs are very limited, with an increased respiratory rate and minute volume, normal cardiovascular findings, and possibly peripheral tetany.

It is vital to exclude other causes of respiratory distress, such as metabolic acidosis and pneumonia, which may cause severe breathlessness with few physical signs. If there is any doubt about the diagnosis, it is essential to perform a thorough series of investigations. The single most useful test is probably ABGs, since normal bicarbonate, low P_{CO_2}, and slightly high P_{O_2} confirms the diagnosis and excludes other pulmonary pathology. Hyperventilation is a diagnosis of exclusion.

Treatment consists of reassurance, explaining the reasons for the patient's symptoms, and offering advice on how to control further episodes.

HIV-related respiratory emergencies

In most places, the majority of HIV-positive patients who are seen in A&E departments actually present with something unrelated to their HIV status, and both patient and doctor may be unaware of that status. Asthma, pneumonia, and pneumothoraces are as common or more common in patients with HIV than in the normal population.

New diagnoses of AIDS are sometimes made in patients presenting with respiratory or neurological symptoms. The major pulmonary infectious agents in patients with AIDS are *Pneumocystis carinii* and *Mycobacterium tuberculosis*, with less common agents including *Aspergillus*, Cryptococcus, and cytomegalovirus. Maintain a high degree of suspicion and seek expert help if in any doubt.

Further reading

1. British Thoracic Society (1993). Guidelines for the management of asthma. *British Medical Journal*, **306**, 776–82.
2. Miller, A. C. and Harvey, J. E. (1993). Guidelines for the management of spontaneous pneumothorax *British Medical Journal*, **307**, 114–16.
3. Hosker, H. R., Jones, G. M., and Hawkey, P. (1994). Management of community acquired lower respiratory tract infection. *British Medical Journal*, **308**, 701–5.

CHAPTER 5

Gastroenterological emergencies

Upper gastrointestinal bleeding

Patients with upper gastrointestinal (GI) bleeding usually present with haematemesis but the condition should be suspected in any patient with a cardiovascular collapse of unknown cause. Always perform a rectal examination, looking for melaena, in patients with unexplained hypovolaemia or hypotension.

Resuscitation

- Check Airway, Breathing, Circulation; give high-flow oxygen; and insert two wide-bore (e.g. 14G) IV cannulae.
- Take blood for full blood count, cross-match, urea and electrolytes.
- If you suspect hypovolaemia start fluid resuscitation with normal saline or colloid.
- Call the gastroenterology and/or surgical team early.

Remember that patients may lose 30 per cent of their blood volume but still have a normal blood pressure so look for the earlier signs of shock such as poor peripheral perfusion and tachycardia. Hypotension is a pre-terminal sign in hypovolaemia.

Causes of upper GI bleeding

Oesophagus oesophagitis, Mallory–Weiss tear, ulceration, varices, tumour.

Stomach gastritis, peptic ulceration, tumour, varices.
Duodenum duodenitis, peptic ulceration.

Initial resuscitation in the A&E department is identical.

Blood transfusion

After 2 litres of crystalloid or colloid, or less if the patient is anaemic, if there are still any signs of hypovolaemia, warmed blood should be started. It may take up to 1 hour to cross-match blood fully, so if the patient needs blood before that give O-negative blood, which is available immediately, or type specific blood (the patient's blood group but not cross-matched) which takes 10 minutes. The gastroenterology and/or surgical team should have been contacted before blood is required.

Who needs a central line?

Peripheral lines are quicker to insert than central lines, and as they are shorter and usually of larger bore, allow more rapid volume resuscitation. A central line is helpful for monitoring cardiovascular status, particularly in older patients with cardiac disease, but should only be inserted once volume replacement through peripheral lines is under way. The most experienced staff available should insert the central line as in hypovolaemic patients it is difficult and complications are more likely.

Continuing management

Continuing bleeding A proportion of patients with upper GI bleeding will continue to bleed despite resuscitation. It is essential that high-flow oxygen is continued and blood is given to maintain the circulation, and these patients require urgent referral to the surgical and gastroenterology teams. They will need to be taken to the operating theatre for urgent endoscopy. Bleeding may be controlled at endoscopy or surgery may be required.

No continuing bleeding In many patients with haematemesis bleeding will stop spontaneously. Volume resuscitation should be continued until the haemodynamic state is restored to normal, with normal blood pressure, heart rate, and peripheral perfusion; and anaemia should be corrected. These patients should be referred to the gastroenterology team for urgent endoscopy. This will enable a diagnosis to be made, so that surgery can be planned if re-bleeding occurs and it may be possible to inject a vessel to prevent further bleeding. There is no evidence that H_2 antagonists, given acutely, reduce re-bleeding.

'Minor' haematemesis Some patients present having vomited a small amount of blood but without any signs of shock (particularly common with Mallory–Weiss tears). They should have IV access secured, blood taken for fbc, U&E, and group and save, and should be referred to the gastroenterology team for admission. They will require endoscopy to find the source of the bleeding.

Variceal bleeding

You may suspect this diagnosis because of a history of liver disease, a heavy alcohol intake, or a previous variceal bleed. Variceal bleeding may be torrential, but stay calm. Initially assess and treat ABC, give oxygen and start volume replacement through at least two wide-bore IV lines as in any hypovolaemic patient. Take blood for a clotting screen (at least PT, PTTR, and platelets), and glucose as well as fbc, U&E, and cross-match. Call the gastroenterology team immediately.

Endoscopic treatment Patients with liver disease, even those with known varices may present with an upper GI bleed not due to varices. Patients should be referred early to the gastroenterology team for emergency endoscopy. Immediate injection sclerotherapy is the treatment of choice to control variceal bleeding. Vasoactive drugs and balloon tamponade are now only used when there is not an endoscopic service available, able to perform injection sclerotherapy.

Drug treatment Vasoactive drugs should only be used after discussion with the gastroenterology team, and with the patient connected to an ECG monitor. Vasopressin reduces bleeding by causing vasoconstriction, but it is contraindicated in ischaemic heart disease and should be stopped if ischaemic chest pain develops. The dose is 20 units in 100 ml 5 per cent dextrose over 15 minutes. There is ongoing research into other agents, your local gastroenterology team will advise you.

Balloon tamponade Sengstaken–Blakemore or Linton tubes incorporate balloons which compress the varices and stop bleeding in 90 per cent of cases. Never attempt to insert these tubes unless the symptoms of the patient have been discussed with a senior member of the gastroenterology team and you have been taught the technique under supervision. While a patient has the tube *in situ* they must be nursed in an ITU environment.

The balloon should not be inflated for more than 18 hours and re-bleeding is common on balloon deflation, so during this time the patient should be transferred to a centre where specialized treatment is available.

Severe rectal bleeding

Causes of severe rectal bleeding in adults include angiodysplasia, diverticular disease, polyps, tumour, and local trauma but initial management in the A&E department is identical for all causes.

Resuscitation

Rectal bleeding can quickly lead to cardiovascular collapse, therefore initially:

- Check ABC, give high-flow oxygen, and insert at least two wide-bore (e.g. 14G) IV cannulae.
- Take blood for fbc, cross-match, U&E
- If you suspect hypovolaemia start resuscitation with normal saline or colloid.
- Call the surgical team early.

See the preceding section on upper gastrointestinal bleeding for assessment of the patient's haemodynamic status, fluid resuscitation, blood transfusion, and the role of a central venous line.

Severe gastroenteritis

Most patients presenting to an A&E department with diarrhoea have mild gastroenteritis without dehydration, are still able to take oral fluids, and may be discharged home with advice about oral fluid replacement. A small proportion, however, have a severe illness, requiring urgent treatment.

Unless the patient is presenting as part of a major outbreak the organism responsible will not be known, but this will not affect the initial management.

Severe dehydration With severe diarrhoea large amounts of fluid and electrolytes, particularly potassium, may be lost resulting in dehydration and hypokalaemia.

Resuscitation Patients with gastroenteritis may present in extremis, severely dehydrated, with cardiovascular collapse, and anuria. As in all critically ill patients the airway must be open and protected, high-flow oxygen must be given and wide-bore IV access should be obtained.

Investigations Blood should be taken for fbc, urea and electrolytes, glucose, and blood cultures, and faeces sent for microbiological examination. If the patient has been abroad let the laboratory known.

An abdominal X-ray should be obtained to look for gross bowel distension and an erect chest X-ray, if the patient is well enough, for evidence of free subphrenic gas, implying bowel perforation.

Fluid resuscitation should be commenced and a urethral catheter placed to allow urine output to be assessed. Normal saline may be used as the initial resuscitation fluid. It is wise not to add potassium to the fluid until the serum potassium is known or the patient is known to be passing urine.

Large amounts of fluid may be required and, particularly in those patients with cardiac disease, central venous pressure monitoring may be required. The placing of a central venous catheter in a dehydrated patient with relatively collapsed central veins is difficult and carries a higher risk of complications than in a well-hydrated patient. This procedure should therefore only be done by experienced staff, once initial resuscitation is under way through a peripheral line.

Analgesia may be required but there is no place for opiates, or any other agent, to reduce stool frequency.

Septic shock may occur in patients with gastroenteritis. If you suspect this, begin resuscitation as outlined above, with oxygen and fluids, but also, once blood cultures have been taken, start broad-spectrum antibiotics. The in-patient/ITU team should be consulted about the choice of antibiotics (see Chapter 12).

Admission Severely ill patients require admission to a high-dependency unit or the ITU. If your local infectious diseases unit cannot provide this level of nursing care the patient may need to be admitted to the general ITU or high-dependency unit, but they will need to be nursed in a side ward to prevent cross-infection.

Acute hepatic failure

Aetiology

The incidence of the various causes of acute liver failure varies around the world, but in the United Kingdom the most common causes are paracetamol overdose and viral hepatitis. Idiosyncratic reactions to drugs or toxins, and other conditions, such as Wilson's disease, malignancy, and auto-immune, or ischaemic liver disease, may cause liver failure.

Classification of hepatic encephalopathy

Grade 1 Mild confusion, slurred speech.
Grade 2 Drowsy but confused and disorientated, inappropriate behaviour.
Grade 3 Sleeps most of the time, incoherent speech, severe confusion.
Grade 4 Unrousable, may or may not respond to painful stimuli.
Survival rates fall with increasing severity of encephalopathy.

Diagnosis

In acute liver failure the stigmata of chronic liver disease, such as spider naevi, ascites, and splenomegaly are likely to be absent, although there may be detectable hepatic fetor or jaundice. The diagnosis is made by obtaining a good history, having a high index of suspicion, and confirming the suspicions with laboratory tests.

Investigation

Many investigations will be performed on patients with acute liver failure, this is not an exhaustive list but contains only those tests that may affect initial management.

- *Blood glucose.* Repeat at least every 4 hours.
- *Urea and electrolytes.* Acute renal failure may occur and will need prompt treatment. There may be profound electrolyte disturbances, see section below.
- *Arterial blood gases.* These are essential for monitoring respiratory function.
- *Full blood count and coagulation studies.* The INR is a good indicator of liver function.
- *Liver function tests*, albumin, magnesium, and phosphate.
- *Other blood tests.* Paracetamol level (may be normal in patients who present late), hepatitis serology for hepatitis A, B, and C, blood cultures.
- *Chest X-ray*, 12-lead ECG, sputum, and urine culture.

Management

Airway   Although usually well maintained until late, the airway may be compromised due to a reduced level of consciousness. It must be kept open and protected.

Breathing   Hypoxia due to intrapulmonary shunting is common, non-cardiac pulmonary oedema, sepsis, and aspiration may also lead to respiratory failure. Patients with normal arterial oxygen tension may have end organ hypoxia due to vascular changes, therefore all ill patients need oxygen by facemask. If hypoxia persists this is an indication for urgent referral to ITU for anaesthesia, intubation, and ventilation.

Circulation   There may be circulatory changes similar to those seen in sepsis, with a reduction of systemic vascular resistance, arterial pressure, and oxygen delivery. Urgent referral to ITU for inotropic support with invasive monitoring is required.

Encephalopathy   The grading of encephalopathy is shown above. Factors, such as hypoglycaemia, hypoxia, electrolyte, and acid-base disturbance, that may contribute to impaired conscious level must be sought and corrected. Sedative drugs should be avoided as they prevent monitoring of disease progression and may have prolonged action. Cerebral oedema is common in acute liver failure. The resulting increase in intracranial pressure reduces cerebral blood flow and may lead to coning. Patients are nursed in a 30 degree head-up position, and, after discussion with the ITU or liver unit, treated with 1 g/kg mannitol, given as a 20 per cent solution. Hourly urine output measurement and careful fluid balance is essential. Fitting must be promptly treated with diazepam or phenytoin.

Metabolic disturbance and renal failure Hyponatraemia is common but does not imply body sodium depletion. Rather it is indicative of water and sodium overload, and should not be treated with sodium supplements. Hypokalaemia should be corrected (with caution in the presence of renal failure) as should hypoglycaemia, hypomagnesaemia, and hypophosphataemia. Renal failure in liver failure may be due to hypoperfusion or hepatorenal syndrome. Every effort should be made to prevent acute tubular necrosis by maintaining blood pressure and renal perfusion. Established acute renal failure, carries a high mortality and such patients must be managed on an ITU or liver unit.

Sepsis This is the most common cause of death in patients with acute liver failure. Prevention and early aggressive treatment are essential. Parenteral broad-spectrum antibiotics (and often antifungals) are given at the first sign of infection and enteral gut decontamination may be performed. The antibacterial/antifungal regime should be decided by the liver unit that will receive the patient.

Clotting problems Bleeding is common in liver failure. Synthesis of clotting factors is impaired, platelet numbers and function are reduced, and DIC may occur. Prolongation of the INR may be used as a prognostic indicator or to monitor disease progression in acute liver failure. Vitamin K should be given and H_2 antagonists reduce the incidence of gastrointestinal bleeding. Active bleeding is treated with blood transfusion, platelets, and fresh-frozen plasma (FFP). Fresh-frozen plasma and platelets may also be given to reverse the coagulopathy prior to procedures such as central line insertion.

Where should the patient be admitted?

Because acute liver failure is an uncommon condition few non-specialists have a great deal of experience in its management. Supraregional centres have developed which specialize in the management of acute liver failure and have access to all modalities of treatment including transplantation. This has been associated with improved survival rates. Patients presenting to the A&E department with acute liver failure should be referred to the hospital's own medical or gastroenterology team, and the ITU physicians, but consideration should be given to their early discussion with and transfer to the supraregional liver centre. Transfer of these severely ill patient must be performed by senior medical and nursing staff experienced in interhospital transfers.

Further reading

1. Williams, S. G. J. and Westaby, D. (1994). Management of variceal haemorrhage. *British Medical Journal*, **308**, 1213–16.
2. Oh, T. E. (1990). *Intensive care manual*. Butterworths, Sydney.
3. Williams, R. (1994). New directions in acute liver failure. *Journal of the Royal College of Physicians, London*, **28**, 552–9.

CHAPTER 6

Neurological emergencies

The unconscious patient

Coma or unconsciousness represents the severe end of a continuum of higher brain dysfunction from normal through to complete unresponsiveness. An objective assessment of the conscious level can be made using the Glasgow Coma Scale (GCS); see Table 6.1. Coma is defined as a GCS less than or equal to 8/15. The GCS should be determined in all patients with a neurological problem.

Profound brain dysfunction carries a risk of ongoing neurological damage, and of problems in other systems, such as airway obstruction, aspiration pneumonia, pressure sores. rhabdomyolysis, dehydration, and renal failure. Supportive care is essential for all comatose patients, and treatable causes of unconsciousness must be identified rapidly.

The GCS is calculated by adding the numbers corresponding to the best response obtained in any limb at any one time. The minimum possible score is 3/15 (completely unresponsive), and the maximum is 15/15 (normal). The standard pain stimuli are digital nail pressure, supraorbital pressure, and sternal pressure. The GCS should be recorded with the indi-

Table 6.1 • The Glascow Coma Scale

E = eye opening	M = best motor response	V = verbal response
4. Spontaneous	6. Obeys commands	5. Orientated
3. To verbal stimuli	5. Localizes to pain	4. Confused
2. To painful stimuli	4. Withdrawal	3. Inappropriate words
1. None	3. Abnormal flexion	2. Incomprehensible
	2. Extension	sounds
	1. None	1. None

vidual components as well as the total score. Thus a patient who opens his eyes and groans when spoken to, localizes sternal pain with his left hand but does not move his right arm would be recorded as GCS 10/15 (E3 M5 V2)

Immediate assessment and management

Airway An obstructed airway causes hypoxia and coma, whereas coma puts the airway at risk. Check for a patent airway and open the airway by whatever means are necessary. Assess for cardiac arrest.

Breathing If the patient is not breathing, begin ventilation with mouth to mask or bag-valve-mask depending on the available equipment. If they are breathing, give high-flow oxygen by facemask and assess for hypoxia.

Circulation Assess for shock which may cause coma. Obtain IV access.

Neurology Quickly assess the patient's conscious state. Is the patient responding to voice or to pain? Are they moving all limbs? Are they vocalizing? Calculate and record the GCS.

Check blood glucose Use glucose sticks or equivalent. Treat hypoglycaemia (< 4 mmol/1) with 50 ml 50 per cent dextrose IV. Do not give this routinely in unconscious patients but wait for the glucose sticks result—it only takes 30 seconds. Hyperglycaemia is bad for the hypoxic or nutritionally deficient brain. If there is doubt about the nutritional status (e.g. alcoholics, anorectics), give 100 mg IV thiamine at the same time.

Anticonvulsants If the patient is fitting or has evidence of localized seizures, start anticonvulsant therapy. Begin with 0.1 mg/kg IV diazepam. (See also the section on status epilepticus, p. 81.)

Naloxone Check for signs of narcotic intoxication. Look particularly for needle marks, pin-point pupils, and low respiratory rate. If there is any doubt, give naloxone 0.8 mg IV (see Chapter 9).

Consider the use of IV thiamine If the history is suggestive of alcohol or profound nutritional deficiencies then the Wernicke–Korsakoff syndrome is a possibility. If in doubt, give 100 mg IV thiamine.

Check the patient's temperature Could hypo- or hyperthermia be the cause?

Formal examination This needs to be thorough, involving not only the neurological system but all others, since disease in many systems may cause coma.

Recheck for any reversible cause It is vital to ensure that there is no ongoing brain damage that could be prevented by a simple treatment. Some doctors use a mnemonic to run through the possible causes. An example of one such system is shown in Table 6.2. Whatever system you use, make sure it is thorough and when under pressure, you can remember it.

Investigation of the unconscious patient

The history and examination should localize the likely cause of unconsciousness. In particular, you should decide if you are seeking an intracranial or extracranial cause. Profound disease in any body system may cause unconsciousness, but the signs should direct you to the relevant system.

The principal investigations for intracranial problems are computed tomography (CT) scan and lumbar puncture (LP). CT scanning can distinguish between subarachnoid haemorrhage, intracranial haemorrhage, and cerebral infarction, and, importantly, can identify neurosurgically correctable lesions. When the patient is a potential candidate for neurosurgery (i.e. relatively young, good premorbid quality of life, and an intracranial lesion is suspected), then urgent CT scanning must be arranged. Remember that the patient must be resuscitated and have a secure, protected airway prior to leaving the resuscitation room for the CT scanner.

Lumbar puncture should not be performed on a comatose patient without first excluding raised intracranial pressure. If meningitis or encephalitis is thought to be the cause of coma, treat with antibiotics prior to performing a CT scan.

Table 6.2 • Causes of unconsciousness or acute confusion

This is one example of a mnemonic for remembering the causes of unconsciousness:

AEIOU FFIITTSS (Vowels and FITS)

Alcohol
Intoxication
Withdrawal (seizures, delerium)[a]
Wernicke[a] Korsakoff syndrome[a]

Epilepsy
Partial seizures[a]
Generalized seizures (inc. non-convulsive)[a]
Post-ictal state

Insulin–Hypoglycaemia
Diabetic 'hypo'[a]
Drug overdose (e.g. chloroquine)[a]
Alcoholic

Overdose
Narcotics[a]
Hallucinogens
Prescription medications

Uraemia–Metabolic
Renal failure (rare)
Hyponatraemia[a]
Hypercalcaemia[a]

Fever–Temperature
Heat stroke[a]
Hyperthermia[a]
Hypothermia[a]

Infection
Bacterial (meningitis, cerebral abscess)[a]
Viral (meningitis, encephalitis)[a]
Extracranial (sepsis)[a]

Ischaemia or Hypoxia
Transient ischaemic attack (TIA)
Pulmonary causes of hypoxia[a]
Post-hypoxic state (e.g. cardiac arrest, strangulation

Trauma
Haematomata (subdural, extradural, intracerebral)[b]
Diffuse damage[b]
Raised intracranial pressure[b]

Tumour
Primary (often with bleeding)
Secondary
Non-metastatic effects (e.g. metabolic)

Stroke = Cerebrovascular accident
Infarct
Haemorrhagic infarct
Subarachnoid haemorrhage

Shock
Hypovolaemic[a]
Cardiogenic[a]
Anaphylactic[a]

Functional–Psychiatric = Diagnosis of exclusion
Hysterical pseudocoma
Catatonic state
Fugue state

[a] Require urgent specific treatment started in the A&E department.
[b] Require urgent review by appropriate specialist team.

Once the ABC and steps listed above under 'immediate assessment and management' have been attended to, if the patient remains unconscious, the common extracranial causes are drug overdose and metabolic abnormality. Specific drug levels are not a particularly useful investigation, since treatment is largely supportive. However, if there is a possibility of multiple drug overdose, screening tests for paracetamol and salicylate should be undertaken. Other tests that should be performed include fbc, U&E, creatinine, LFTs, calcium, chest X-ray, and ECG.

Disposition

All patients who remain in coma after resuscitation must be reviewed urgently by the medical and ITU teams and admitted to an ITU for airway management and further investigation.

Acute confusion

Acute confusion is the less severe end of the spectrum of higher brain dysfunction. The causes and investigation follow much the same pattern as those for the comatose patient, but the therapy is generally simpler.

The essence of acute confusion is disorientation in time, person, or place, in association with a decrease in conscious level. This can be difficult to identify, particularly in a patient with pre-existing chronic confusion and those with language difficulties.

Always obtain as much history as possible (the patient may not be much help) and seek an opinion on the patient's mental state from someone who knows them and is fluent in the patient's first language. If in doubt, investigate and treat as acute confusion.

The common causes of confusion in the A&E department setting vary with age. Young, previously healthy people most often present confused in the setting of a drug overdose, particularly anticholinergics and hallucinogens (Chapter 9). Only in the most severe cases do sepsis, hypoxia, or shock lead to impaired cerebral function. On the other hand, elderly patients, with pre-existing cerebrovascular disease, are likely to become confused in response to mild physiological insults, such as chest infections, urinary tract infections, or small cerebrovascular events.

Immediate assessment and management

1. Airway, Breathing, Circulation.

2. Seek and treat any reversible cause not dealt with by ABC:
- hypoglycaemia (perform blood glucose),
- narcotic intoxication,
- Wernicke–Korsakoff syndrome,
- hypo- or hyperthermia.

3. Establish IV line and take blood for investigations.

4. Begin formal assessment with thorough history and examination.

Investigation

All acutely confused patients with no definite cause evident on history and examination require further investigation with full blood count, urea and electrolytes, creatinine, calcium, LFTs, blood cultures if febrile, arterial blood gas, midstream urine, chest X-ray, and ECG. Cranial computed tomography and lumbar puncture should be reserved for those with clinical evidence of intracranial pathology.

Disposition

All acutely confused patients, except those whose problems are completely resolved in the A&E department (e.g. hypoglycaemia), must be admitted by the medical team.

Status epilepticus

Status epilepticus is defined as continual seizures lasting over 10 minutes or recurrent seizures without full recovery between episodes. This is a medical emergency because of the risk of permanent cerebral damage. Initially, this risk is due to hypoxia, acidosis, and the risk of hyperthermia, but prolonged fitting is directly neurotoxic. Drug therapy for status epilepticus follows a stepwise progression, advancing to the next step if control is not obtained, as listed in Table 6.3.

Immediate assessment and management

Cardiac arrest may present with loss of consciousness associated with epileptiform movements. Always check for the presence of a central pulse.

Airway Check that the airway is clear, and, if obstructed, clear by appropriate means. Inserting an oropharyngeal airway

Table 6.3 • Treatment of seizures

Step	Drug	Comments
1	Diazepam 10 mg or midazolam 5 mg	Repeat once if necessary. The dose should be reduced in the elderly
2	Phenytoin 18 mg/kg over 20 minutes *or*	NB: caution if the patient is on phenytoin.
	paraldehyde	10 ml rectally or IM, 5 ml into each buttock
3	**Get expert help** IV clonazepam or diazepam infusion or thiopentone, and ventilation	These drugs cause severe respiratory depression and should only be used by senior, experienced staff

into a fitting patient can be difficult or impossible. If the patient is maintaining their airway it may be better to stop the fitting before attempting to insert an oropharyngeal airway.

Breathing If the patient is breathing give high-flow oxygen through a facemask, if not breathing then ventilate with bag-valve-mask, ensuring the airway remains clear.

Circulation If there are no palpable pulses, commence cardiopulmonary resuscitation and treat as for cardiac arrest. If pulses are palpable, insert an IV cannula and take blood from it. If possible avoid the flexures for IV cannulation in a fitting patient.

Seek and start treatment of any easily reversible cause

- *Hypoglycaemia* Immediate bedside and formal blood glucose levels are mandatory in all patients with seizures.
- *Hypoxia* Treat by the ABC manoeuvres described above. Check the SaO_2 with a pulse oximeter. The SaO_2 ids likely to remain low while the patient is fitting.
- *Malignant hypertension* Taking the blood pressure is difficult in a fitting patient but, if achieved, a grossly elevated BP (> 220/130) should alert you to this possibility. Obtain expert help in management
- *Eclampsia* In a female patient, check for pregnancy. If the patient is pregnant contact an obstetrician immediately as you manage ABC and treat the seizures. The definitive and urgent therapy is to deliver the baby.

Begin specific anticonvulsant therapy Start IV diazepam, titrated in 2 mg aliquots, to a total of 10 mg. Wait 5 minutes and, if there is no effect, repeat the dose. If IV access is impossible, give twice the dose rectally and call for senior help. If the patient continues to fit begin a loading phenytoin infusion. If the patient is not already taking phenytoin, give 18 mg/kg at no more than 50 mg/minute. If the patient is taking phenytoin, and you are certain they do not have phenytoin toxicity then give a 'top-up' dose of 5–10 mg/kg. If the seizures are still not controlled the next steps are IM paraldehyde, a clonazepam or diazepam infusion, or thiopentone and intubation (see Table 6.3). Seek expert help.

Formal history and examination The majority of cases of status epilepticus are patients with known epilepsy, with a precipitating cause such as non-compliance, infection, or a change in drug metabolism (e.g. drug interactions). Previously healthy patients with status epilepticus must be fully examined and then investigated urgently. In the young, the likely causes are stimulant drug overdose (e.g. mefanamic acid, theophylline, cocaine, tricyclics), drug withdrawal (alcohol, opioids or benzodiazepines), and intracranial infection. In the old, think particularly of cerebral tumour, stroke, electrolyte disturbance, and hypoxic encephalopathy. However, any of the causes of coma (see Table 6.1) can also cause seizures.

Investigation

The degree of investigation depends on the likely cause and response to therapy.

Known epileptics who respond rapidly to treatment, and who recover to normal within 1–2 hours need only a simple screen for causes not evident on clinical examination: fbc, U&E, creatinine, calcium, ABG, anticonvulsant levels, chest X-ray, MSU, any relevant cultures.

Previously healthy patients who present with their first seizure in status epileptics need a more complete work-up, including a CT scan, seeking structural lesions of the brain.

Any patient who does not respond rapidly to the treatment regimen will require ongoing investigation as part of their management. As treatment progresses seek expert advice and obtain a CT scan; lumbar puncture may be indicated after the patient is reviewed by a senior member of staff, if there is no contraindication. Electrocardiogram, electroencephalogram, toxicology screen (theophylline, isoniazid, chloroquine, carbon monoxide, cyanide, strychnine (which causes muscle spasm, not true seizures)), and echocardiogram (seeking endocarditis) may also be required.

Disposition

All patients with status epilepticus must be admitted to a medical unit. All patients who did not respond to initial therapy must be admitted to an ITU.

Single seizure

Patients often present to A&E departments after a single seizure. Some have a fit while in the hospital or A&E department for other reasons. The two most common causes are alcohol withdrawal and non-compliance in known epileptics.

If a patient is seen fitting, begin treatment along the lines of status epilepticus as above. The fit will usually stop within 3 minutes, and if it does not, then continue with treatment as above.

When examining a patient after a single seizure, perform a thorough neurological history and examination. In particular, determine whether the seizure had any atypical features. 'Typical' generalized seizures start with an aura, followed by a rigid tonic phase, often with a loud cry, followed by generalized clonic movements, often with urinary incontinence, lasting less than 3 minutes. This is followed by a post-ictal phase of slowly lightening confusion lasting less than an hour. Absence of one of these features does not make the fit 'atypical' but absence of many, prolonged fitting, or any focal signs before, during or after the seizure does.

In addition to the nature of the particular seizure, find out how it relates to the patients' usual pattern of fits if they are epileptic. A second seizure in two days in a patient who normally has less than one per year is much more significant than a second seizure in two days in an alcoholic who has regular withdrawal seizures.

Criteria for discharge of a patient after a single seizure:
1. stable home environment with someone able to help in the event of recurrence, and follow-up arranged *and*,
2. normal examination after observation *and*,
3. no atypical features, particularly focal neurological features *and*,
4. either:
(a) In a known epileptic, a seizure that is in keeping with their usual pattern of seizures with no precipitating illness apparent on clinical examination. Consider taking an anticonvulsant level, and arranging for the patient to be reviewed by their GP with the result. If non-compliance is the problem, give the patient their usual medication before discharge.

or

(b) Controversial: some neurologists are happy for patients presenting with their first seizure with no apparent precipitating cause to have further investigations as an out-patient. If this is your local policy make arrangements for electroencephalogram, CT scan, and neurological follow-up.

All other patients should be admitted after a seizure.

Todd's palsy

After a focal seizure, including one with a secondary generalized convulsion, some patients experience a period of dysfunction of the part of the cerebral cortex from which it arose. This manifests as a reversible motor weakness of the relevant body area, lasting longer than the post-ictal phase, but less than 24 hours. At times it may be alarming if it involves speech or swallowing. Treatment is simply supportive, but airway protection may rarely be required. If it has not occurred before, further investigation as for a focal neurological deficit (p. 93) is required.

Pseudoseizures

A small number of patients present feigning fits. It may be obvious that the limb movements are voluntary, or that the patient is conscious, but in some patients the pseudoseizures closely resemble epileptiform seizures. Two observations that may be of benefit are that during a *grand mal* seizure the pupils usually dilate and the SaO_2 is almost always low. If you are in any doubt about the diagnosis treat the patient as if it is a genuine fit. Do not risk leaving a patient in untreated status. If the glucose sticks and the SaO_2 are normal and you are certain that it is a pseudoseizure confront the patient and tell them you know they are not having a true fit. This usually results in an almost immediate recovery.

Headache

The important factor in assessing a patient with headache is to exclude the serious, potentially life-threatening causes. Pay particular attention to any patient with a new headache, that is, a headache with characteristics unlike those experienced before.

The common headache syndromes are shown in Table 6.4. A history of sudden, dramatic onset should alert you to the possibility of subarachnoid haemorrhage (SAH). Remember that a significant proportion of patients disabled by SAH have previously sought medical care with a 'warning bleed' (i.e. a grade I SAH). Bilateral headache with fever and/or neck stiffness and/or drowsiness should alert you to the possibility of meningitis or encephalitis. A headache of gradual onset, worse in the mornings, worse on straining and accompanied by vomiting suggests raised intracranial pressure. Gradual onset of new headaches in patients over 55 years suggests temporal arteritis. Nonetheless, the common headaches seen in A&E departments are classical migraine, tension headache, and pain having its origin in the sinuses or the neck. It is also relatively common to see patients with high fever and secondary headache due to tonsillitis, urinary tract infection, or other infection, but a high degree of suspicion of meningitis must be maintained.

A thorough history and examination must be performed on all patients presenting with headache.

The first four syndromes listed in Table 6.4 are the ones that must be excluded in the emergency setting. In addition to sinusitis and otitis media, other infections, such as tonsillitis,

Table 6.4 • Headache syndromes

Syndrome	Headache	Pattern	Associated features
Meningitis	Bilateral, severe	Steadily worsening	Fever, neck stiffness, drowsiness, photophobia. Later signs of septic shock
Subarachnoid haemorrhage (SAH)	Bilateral, often occipital	Sudden dramatic onset, often on straining	Variable: from none through neck stiffness and vomiting to profound coma
Raised intracranial pressure	Bilateral	Subacute onset worse on straining, worse in mornings	Vomiting, blurred vision
Temporal arteritis	Unilateral or bilateral	Subacute onset	Tender temporal arteries, elderly patient. Jaw claudication, sudden blindness, high erythrocyte sedimentation rate (ESR)
Migraine	Unilateral throbbing	Acute recurrent sometimes triggered by food or drugs	Aura (usually visual). Often nausea and vomiting Sometimes focal neurological signs
Cluster headaches	Unilateral usually around eye	Frequent short (< 60 min) attacks for a few weeks	Variant of migraine, treatment similar
Tension headache	Bilateral usually temporal/occipital	Subacute to chronic, worse at night	Related to stress. May occur a few days after the major stress has finished
Sinusitis	Unilateral or bilateral, frontal or facial	Subacute, usually with upper respiratory tract infection	Local tenderness, fever. Fluid level seen on sinus X-ray
Cervical spondylosis	Unilateral, usually Occipital	Variable pattern	Exacerbated by movement, cervical spondylosis on X-ray
Otitis media	Unilateral or bilateral	Subacute	Otitis media on ear examination

dental abscess, and even dental caries may present as headache.

Investigation

Investigation depends on the results of the history and examination. If there is any suspicion of SAH then the patient requires a CT scan and lumbar puncture. If there is any suspicion of meningitis, then full investigations are required, as described below. If there is a suspicion of raised intracranial pressure, then refer for investigation. If the clinical diagnosis is definitely that of a simple headache, then no tests are warranted.

Treatment

Treatment of simple headache is best accomplished with simple analgesics rather than narcotics. Use a parenteral anti-emetic, such as metoclopramide 10 mg IV/IM, or chlorpromazine 25 mg IM if a sedative effect is required, followed by oral analgesics, such as aspirin (e.g. 3 × 600 mg aspirin/ 8 mg codeine tablets). Place the patient in a quiet and darkened area and review in 30 minutes. Patients who do not respond to this simple analgesic approach should be reviewed by the medical team.

For patients with migraine, specific serotonin antagonists are now available. Use sumatryptan 6 mg subcutaneously or 100 mg orally. Again, a quiet, darkened area is appropriate.

Disposition

Any patient with significant or disabling headache which cannot be relieved in the A&E department should be referred to the medical team for review, no matter what diagnosis is being considered.

Meningitis

Bacterial meningitis is a life-threatening medical emergency. Viral meningitis is more common, and usually presents with a gradual course, allowing more time for diagnosis. However, bacterial meningitis may present on a background of viral illness, and will tend to have a subacute course if the patient has been treated with any antibiotics in the community. Meningitis is characterized by headache, fever, neck stiffness, meningeal irritation (Kernig's sign), and signs of systemic infection. As the disease progresses, drowsiness, confusion, and coma occur. Depending on the organism and the course, the patient may be overwhelmed by systemic sepsis, or may die a 'neurological death' without septic shock.

Because meningitis is so serious, a low threshold for diagnosis should be maintained. It is expected that some lumbar punctures performed for suspicion of meningitis will be normal. If you suspect meningococcal meningitis then antibiotic therapy, benzyl penicillin 2.4 g IV, must be given immediately. It is better to have a live patient than a good microbiological diagnosis. Early antibiotic therapy is the most important factor in determining outcome in adult meningitis. Patients with septic shock need fluid resuscitation, but bear in mind that most units prefer to maintain patients with meningitis a little 'dry' in order to reduce cerebral oedema. A guide to interpreting cerebrospinal fluid (CSF) results is shown in Table 6.5.

Immediate assessment and management

Airway If the airway is obstructed clear it as necessary.

Breathing If not breathing then ventilate using bag-valve-facemask connected to high-flow oxygen. If breathing, give high-flow oxygen through facemask.

Circulation If no cardiac output commence cardiopulmonary resuscitation and treat as cardiac arrest. Insert an IV line, take blood for fbc, U&E, glucose, blood cultures, and commence IV fluids. If in shock begin resuscitation with IV colloid such as gelofusin.

Neurological assessment Assess degree of consciousness (Glasgow Coma Scale) and assess for focal neurological deficit. Look for signs of meningism such as Kernig's sign. Quickly examine the fundi if the patient will allow it.

Rapid history and examination The more ill the patient, the less time should be wasted before treatment. Nevertheless, look quickly for signs of infection elsewhere (pneumonia, otitis media), hints as to the organisms, particularly rash (meningococcus), and signs of trauma or recent spinal/cranial surgery.

Treatment Give 2.4 g benzyl penicillin IV and 2 g of a third-

Table 6.5 • Guide to cerebrospinal fluid

	Normal	Bacterial meningitis	Partly treated bacterial meningitis	Viral meningitis	Encephalitis	TB Meningitis	SAH
Appearance	Crystal clear	Often turbid	May be turbid	Usually clear	Usually clear	Often fibrin web	Turbid Xanthochromia
White cell count	<10	>500	>50	>50	>50	>10	1/1000 RCC
White cell type	Lymphocytes	Polymorphs	Variable	Mononuclear	Mononuclear	Mononuclear	Polymorphs
Red cell count	<10	<10	<10	<10	<10	<10	>400
Glucose	>2/3 plasma	<2/3 plasma	Variable	>2/3 plasma	>2/3 plasma	<2/3 plasma	>2/3 plasma
Protein	<0.4 g/l	1–5 g/l	1–5 g/l	<1.5 g/l	<1.5 g/l	1–5 g/l	Raised
Pressure	<200 mm CSF	Often raised	Variable	Often raised	Usually normal	Variable	Variable

generation cephalosporin (cefotaxime or ceftriaxone) IV. If venous access is not obtainable rapidly in a very ill patient, use the IM route rather than wasting time. If there is any possibility of herpes simplex encephalitis, particularly temporal lobe signs, personality change, CT findings or LP findings then give acyclovir 10 mg/kg IV.

If you suspect meningitis, the patient is drowsy or comatose, or has hypotension, signs of severe sepsis or meningococcaemia, then proceed directly to treatment with antibiotics. Do not delay treatment to perform a lumbar puncture.

Thorough assessment Finish obtaining a more thorough history and examination, obtain chest X-ray and any other indicated tests, including CT scanning if any focal signs or coma.

Disposition

Arrange admission to a medical unit. If the patient has an abnormal conscious level then they should be reviewed for admission to the ITU. Ensure that arrangements are made to give prophylaxis, if indicated, to the patient's contacts. This is normally done in association with a microbiologist. Offer prophylaxis to intimate (very close, e.g. kissing) contacts of patients with meningococcal meningitis, and intimate contacts aged less than 5 years for *Haemophilus* meningitis. Give rifampicin 600 mg twice daily for 2 days in adults; 10 mg/kg twice daily for 2 days in children aged more than 1 year; 5 mg/kg twice daily for 2 days in children aged less than 1 year. Warn about orange staining of urine and interference with oral contraceptives.

Patients with viral meningitis are also admitted, although management is supportive. Despite the course given above, some lumbar punctures will be performed on patients with a relatively low suspicion of bacterial meningitis. If the CSF obtained on LP is clinically crystal clear, and the patient is well, then wait for the microbiology results before proceeding. If there is any doubt or delay, give antibiotics immediately after LP.

Patients with normal CSF findings or a 'bloody tap' (CSF contaminated with blood) alone (with SAH excluded by xanthochromia studies) may be discharged after a short rest provided that a cause has been established for their symptoms and treatment started as appropriate. Some hospitals have specific protocols on management or admission after lumbar puncture which should be observed. Whilst vigorous exercise should be avoided it is not normally necessary to enforce strict bed rest.

Note that there is considerable overlap between viral meningitis, encephalitis, and partially treated bacterial meningitis— these are so-called 'aseptic meningitis'. CSF with these findings should have further microbiological investigations for other causes, such as India Ink microscopy for *Cryptococcus neoformans*, and specific serology for syphilis and viruses. Seek microbiological advice. Bacterial antigen may be detected in the CSF where there is no growth on culture.

Subarachnoid haemorrhage

Subarachnoid haemorrhage (SAH) usually occurs from small berry aneurysms located around the Circle of Willis, sometimes from AV malformations, and in 20 per cent of cases without a site identified. It may occur at any age. The classical presentation is of sudden onset of headache, with or without neck stiffness, vomiting, or photophobia. The presenting symptoms may range from mild headache of sudden onset through to catastrophic collapse and brain death. In severe cases, it is diagnosed as part of the work-up of coma. SAH is graded clinically:

I. Meningism, no neurological signs.
II. Normal consciousness, cranial nerve signs.
III. Drowsy with any focal signs.
IV. Drowsy with major deficit (e.g., hemiplegia).
V. Moribund, deep coma.

This is the information required by the neurosurgeon, as prognosis decreases rapidly particularly above grade II. However, emergency management follows basic principles.

Immediate assessment and management

Airway If obstructed clear by appropriate means.
Breathing If not breathing, ventilate with self-inflating bag and prepare for tracheal intubation. If breathing, give high-flow oxygen through facemask.

Circulation Although hypertension may contribute to SAH, remember that hypertension and bradycardia are late signs of raised intracranial pressure and should not be treated with cardiovascular medication. SAH may cause quite profound ECG changes. Obtain IV access and take blood for fbc, U&Es, blood glucose.

Neurology Perform a thorough neurological examination seeking particularly meningism and localizing signs.

Assessment Formal history, examination, and any other investigations (e.g. chest X-ray, ECG), as warranted.

Arrange a diagnostic procedure CT scanning is 95 per cent sensitive to SAH, and should ideally be performed on all patients in whom this diagnosis is suspected. However, an LP is necessary in those patients with the clinical suspicion of SAH and a normal CT. If the patient is alert (GCS 14/15 or 15/15) and has no focal neurological signs, and there is no CT scanner available, then an LP may be performed as the primary diagnostic procedure. In cases where the lumbar puncture is contaminated (a 'bloody tap') seek the advice of your hospital's laboratory on cell counts and xanthochromia studies.

Disposition

All patients with SAH proven on CT or LP require admission to a neurosurgical unit. The only exception may be elderly patients with grade IV or V lesions, who have such a poor prognosis that they may be admitted to a medical unit for terminal care. Patients with normal LP findings may be discharged if another cause for their headache is identified and treated.

Sudden focal neurological deficit

This is the normal presentation of a 'stroke' (cerebrovascular accident; CVA) and sometimes a transient ischaemic attack (TIA) which resolves within 24 hours, both due to cerebrovascular disease. This terminology is non-specific, since it is not clinically possible to distinguish between infarction and haemorrhage, nor to be certain of the site of the lesion in many cases. Although stable patients with neurological deficits tend to be given low priority in the A&E department because there is currently no specific therapy available, the condition does represent an emergency with brain tissue at risk.

In younger patients without a high risk of cerebrovascular disease, consider subarachnoid haemorrhage, haemorrhage into a cerebral tumour, embolism from atrial fibrillation or bacterial endocarditis, or multiple sclerosis as causes of neurological deficit. If the picture is of a transient deficit, consider focal epilepsy or atypical migraine.

Immediate assessment and management

1. Airway, Breathing, Circulation.

2. Perform neurological assessment. Is the deficit unilateral (cerebrum) or bilateral (brainstem)? Which limbs are most affected? Is it sensory, motor, or mixed? Are there associated visual field defects or cortical function deficits? What is the time course? Is it improving or worsening? Depending on the site of the lesion it may be possible to localize clinically to the brainstem, internal capsule, or a particular area of cortex.

3. Consider whether there is any treatable cause:

- *Hypoglycaemia* In any patient, but particularly those with cerebrovascular disease, hypoglycaemia may present as a focal deficit such as hemiplegia.
- *Extradural, subdural haematomata* Patients with head injury may present with a syndrome indistinguishable from a stroke.
- *Atrial fibrillation (AF)* Following CT scan to exclude an intracerebral haemorrhage, patients with AF and resultant emboli strokes will require heparinization to prevent further embolic phenomena. All patients with 'strokes' should have an ECG.
- *Subarachnoid haemorrhage* Consider whether the focal deficit is part of the picture of subarachnoid haemorrhage. If this is a possibility, arrange CT scanning and investigation as described above.
- *Atypical migraine* This usually occurs with headache, and resolves fairly quickly.
- *Seizures* may occur secondary to a stroke, may mimic a focal deficit (although twitching is more common), or may produce a focal deficit (Todd's palsy).
- *Malignant hypertension* is rare but requires urgent treat-

ment if it is causing neurological signs (see Chapter 3). Be suspicious in those with diastolic > 130 mmHg. The blood pressure should be lowered slowly. Mild to moderate elevations in the blood pressure immediately following a stroke should not be treated as this can further reduce cerebral perfusion.

- *Guillain-Barre syndrome* Usually a progressive bilateral motor weakness rather than a 'stroke'.

4. Perform full assessment and examination. Careful examination of cardiovascular and respiratory systems is required in addition to neurological assessment. Insert an IV line, take blood for fbc, U&E, creatinine, and, if the plan is to treat the patient with heparin, coagulation studies. Ensure ECG and chest X-ray are obtained. If the patient will be unable to use a bedpan or a toilet, insert a urinary catheter. If patient has any bulbar involvement, make them 'nil by mouth' and ensure adequate IV hydration. If there is a large neurological deficit, ensure pressure area care begins in the A&E department.

Investigation

Patients with atypical features to their 'stroke', such as youth, seizures, progressive course, or a pattern not explainable by a single anatomical lesion, should have early rather than late investigations. The same is true for those with a suspicion of cerebellar haemorrhage (headache and acute cerebellar signs) since posterior cranial fossa haemorrhage may require neurosurgical intervention. In large hospitals, most patients with 'strokes' have a CT scan at some time during their admission, but this should be considered early in those patients with atypical features.

Disposition

All patients with acute strokes should be admitted under a medical unit. Occasionally, a patient presents late with a minor lesion from a stroke (e.g. mild dysphasia), and is already improving. Such rare cases may be able to be discharged to care of their GP if adequate community resources are available.

Patient with TIAs, which have completely resolved, may be discharged for medical clinic follow-up if they have a suitable home environment and immediately treatable causes as above, have been ruled out. Take fbc, U&E, creatinine, syphillis serology, ECG, and chest X-ray if not previously investigated for vascular disease. Ensure the patient is taking aspirin 100–300 mg/day, if not contraindicated by peptic ulceration or coagulopathy, and is on appropriate antihypertensives if indicated.

Spinal cord compression

Non-traumatic compression of the spinal cord usually occurs secondary to metastatic malignancy in the spinal column. Rarely is it due to spinal or epidural abscess, or primary malignancy. Although many patients with metastatic malignancy are already receiving palliative care, the severity of this condition warrants surgical therapy in all but terminal cases.

Patients present with loss of function (sensory and/or motor) distal to the site of the lesion, and commonly sphincter involvement. There is usually pain in the back at the site, and fever if an epidural abscess is present. There may be complete loss of cord function (as in cord transection) a recognizable pattern (e.g. Brown–Sequard syndrome or lower motor neurone signs at the level, with upper motor neurone signs below), or no recognizable pattern to the symptoms and signs, especially if the cauda equina is involved.

Assessment and management

1. Airway, Breathing, Circulation.
2. Insert IV line, take blood for fbc, U&Es, serum calcium if metastatic malignancy suspected, blood cultures if febrile.
3. Thorough clinical examination documenting level of lesion and function affected, and seeking any underlying pathology, such as primary malignancy.
4. Obtain plain X-ray of the spine including above and below the level suspected.
5. If there is clinical suspicion of an epidural abscess begin empirical antibiotic therapy. Seek advice from your local microbiologist, but broad-spectrum cover is appropriate.
6. If the patient has metastatic malignancy and *no* evidence of infection, discuss with the oncologists whether they wish to start treatment with dexamethasone.

Disposition

Refer the patient to a neurosurgical service. The required investigation is CT and/or myelography or magnetic resonance imaging (MRI). Appropriate advice from the neurosurgeon is necessary depending on availability of facilities. Some patients with malignancy may be admitted under oncologists or radiotherapists.

Guillain–Barré syndrome

This rare disease is an acute demyelinating inflammatory polyradiculopathy, which usually presents as a symmetrical ascending flaccid motor paralysis, commonly with some sensory symptoms. It usually occurs some weeks after a viral illness, but a wide variety of patterns are described, and the course ranges from slow to fulminant. The diagnosis is rarely made in the A&E department, but a specific and effective therapy (plasmapheresis) is available so it is important that the condition is not missed.

In the emergency setting, the most important aspect of the disease is respiratory function. Check formal spirometry on all such patients and, if abnormal, obtain blood gases, preferably breathing room air. Because the lungs are normal, patients with respiratory failure may become markedly hypercarbic without hypoxia if they are given supplemental oxygen without checking respiratory function.

Indications for early ventilation are FVC < 1.5l, P_{CO_2} > 45 mmHg, or any clinical breathing difficult on room air, including inadequate cough. Seek expert help and ITU review.

Bell's palsy

This is not a life-threatening condition, but is an alarming disease to have, and so may present to A&E at times when other services are not available. The presentation is of acute motor weakness of the VIIth cranial nerve (weak unilateral facial muscles), sometimes with pain. The cause is believed to be a viral polyneuropathy which may affect other cranial nerves.

Treatment is symptomatic (artificial tears and dark glasses) awaiting a recovery, but some physicians give prednisolone 40 mg twice daily for 5 days. Seek advice from your medical team and arrange medical or ear, nose, and throat (ENT) clinic follow-up.

Always ensure that there is no local cause within the ear before you diagnose Bell's palsy.

Trigeminal neuralgia

This is another non-life-threatening condition which is alarming and painful for the patient. Commonest in older women, it presents with brief spasms of severe unilateral facial pain, often precipitated by touching a 'trigger point' on the face. There are no specific diagnostic tests, and simple analgesics are rarely sufficient to relieve the symptoms. Start treatment with carbamazepine 100 mg thrice daily, and arrange medical follow-up.

Endocrine emergencies

Diabetic emergencies

Hypoglycaemia

Patients with hypoglycaemia are usually diabetics who present comatose or semi-comatose after a missed meal or unaccustomed exercise. However, be alert for the possibility of hypoglycaemia in any diabetic with neurological signs: confusion, dysphasia, or hemiplegia are possible presentations. Further, all patients with abnormal conscious state require an urgent blood glucose: hypoglycaemia may be seen in alcoholism, anorexia, and after a variety of overdoses.

Immediate assessment and management

1. Airway. If the airway is obstructed, clear it.
2. Breathing. If not breathing, use bag-valve-facemask, if breathing then use high-flow oxygen.
3. Circulation. Insert an IV line and take blood for a blood sugar level.
4. Give glucagon 1 mg IV stat or 25 ml 50 per cent dextrose IV stat. They are both effective in the treatment of diabetics with hypoglycaemia. Dextrose may be more readily available but it is a very viscous solution, requiring a wide-bore cannula, and it causes damage to veins.
5. Consider a nutritional cause for the hypoglycaemia (anorexia, alcoholism). If this is a possibility, give 100 mg of IV thiamine at the same time as the glucose. Glucagon is ineffective in hypoglycaemia due to liver disease or anorexia.

Most diabetics with hypoglycaemia will respond rapidly to IV glucagon or dextrose. They should become awake and alert and be able to give a history of events leading up to the episode, otherwise, asses carefully for other causes of coma (see Chapter 6). If no other cause is identified consider prolonged hypoglycaemia with significant cerebral damage.

Disposition Most diabetics treated for hypoglycaemia can safely be discharged to the care of the family or friends. It is important to ascertain why the episode occurred (e.g. missed meal), and take appropriate steps to ensure there is no recurrence. Indications for admission are persistent neurological signs or abnormal conscious state, inadequate care available at home, or recurrent attacks requiring stabilization of therapy in hospital. Always ensure that follow-up with the patient's general practitioner is arranged.

Many diabetics now treat their own attacks with subcutaneous or intramuscular glucagon injections. This is an effective and rapid treatment for diabetic hypoglycaemia although it should be noted that it is ineffective in patients without hepatic glycogen stores (i.e. nutritional deficiency).

Diabetic ketoacidosis

This is a profound metabolic and circulatory upset, which occurs in type I diabetics who have not received enough insulin. The most common setting is that of a non-compliant, often adolescent diabetic, but it may be precipitated by intercurrent illness, drug interactions (especially steroids), and is still seen as the first presentation in new diabetics. The history is usually of deterioration over a few days. Beware of atypical presentations such as abdominal pain or respiratory difficulty. The major manifestations of diabetic ketoacidosis (DKA) are:

Circulatory shock Hyperglycaemia causes an osmotic diuresis and patients are grossly dehydrated.

Acidosis The ketone bodies acetoacetate and hydroxybutyrate cause a profound metabolic acidosis with secondary hyperventilation. This may be exacerbated by circulatory failure.

Hyperglycaemia The high blood glucose and high osmolarity are not of themselves as dangerous as the other manifestations, but obviously require treatment. There may be ketoacidosis without a greatly elevated blood glucose (see p. 104).

Electrolyte abnormalities Patients with DKA have hyponatraemia, due to sodium loss from the osmotic diuresis, and an additional, artefactual lowering of the measured serum sodium due to the presence of high glucose. In addition, most patients have hyperkalaemia but a significant total body potassium deficit. The latter is particularly important since serum potassium will fall rapidly once fluids are replaced and insulin given.

A possible underlying cause The precipitating cause may or may not be manifest. Unless there is very clear historical evidence of non-compliance, precipitating pathology, such as infection or intra-abdominal pathology, must be excluded.

Immediate assessment and management

Airway If the airway is obstructed open it by whatever means are necessary.

Breathing The patient is likely to be breathing rapidly and deeply. In all ill patients use high-flow oxygen until you have corrected shock and hypoxia. Connect a pulse oximeter.

Circulation The patient will be in shock. Insert a wide-bore line, take blood for fbc, U&Es, blood glucose, and blood cultures. Begin immediate fluid resuscitation with 1000 ml normal saline over 30 minutes. Follow this with a further 1000 ml over approximately 60 minutes. Take blood for blood gases but start treatment before you have the results. Patients can be safely managed by judging the degree of metabolic acidosis by means of venous bicarbonate or venous pH.

Begin insulin therapy Be guided by the protocols in your hospital. Otherwise make-up 50 units of insulin in 50 ml of normal saline and start an IV infusion at 0.1 units/kg/h (e.g. run at 7 ml/h for a 70 kg man). This may be preceded by 0.1 units/kg soluble insulin IV stat.

Seek an underlying cause Obtain a further history from the patient and/or relatives as appropriate. Perform a thorough physical examination, check chest X-ray and midstream urine.

Ensure ongoing management The essential features are:

Continued fluid management until no longer dehydrated. These patients may have a very large fluid requirement (10 litres or more to fully restore their normal state) but only the first 1–2 litres needs to be given quickly. Use normal saline until the blood glucose level falls to below 15 mmol then use 4 per cent dextrose–0.18 per cent saline. While the blood glucose is high, urine output is not an adequate measure of hydration.

Give supplemental potassium From the third litre of IV fluids onwards, add potassium to each bag. This should be titrated to the serum potassium, bearing in mind that it will fall rapidly: 40 mmol potassium per litre of saline is a good starting point. Serum potassium should be measured every hour until the patient is stable.

Ongoing insulin requirements Measure the blood glucose hourly, and halve the insulin dose once the blood glucose has fallen below 15 mmol. If the blood glucose does not fall at around 5 mmol/h then the insulin infusion needs to be increased or decreased as appropriate. Once again, your own hospital's protocol should be consulted. The insulin infusion must be continued until after acidosis is corrected, not ceased when the blood glucose falls.

Disposition All patients with DKA require admission to hospital. Any patient with a pH < 7.20 needs to be considered for ITU therapy.

Cautionary points The place of sodium bicarbonate in the management of metabolic acidosis remains controversial. Some doctors would give 100 mmol of sodium bicarbonate, over 6 hours, to any patient with a pH of < 7.0, whereas others feel that the risks of cerebral oedema, reduced oxygen delivery to the tissues, increased intracellular acidosis, and profound hypokalaemia outweigh the possible benefit.

Repeated measurement of arterial blood gases is rarely essential in the management of DKA, except in patients with respiratory failure or who have received IV sodium bicarbonate. For ongoing assessment of acid–base status, the serum bicarbonate taken on a biochemical profile is normally adequate.

In addition to hyponatraemia and misleading hyperkalaemia, these patients usually have leucocytosis whether or not infection is present. Some creatinine assays cross-react with ketone bodies, so do not consider the creatinine level accurate.

Abdominal pain and ileus are common, so a nasogastric

tube will be of benefit. Serum amylase is frequently raised without overt pancreatitis, but severe pancreatitis can precipitate DKA.

Hyperosmolar non-ketotic coma

Elderly type II diabetics are prone to the insidious development of a hyperosmolar state, with blood sugar levels >50 mmol and serum osmolarity >350 mmol. This is a very serious disease, with up to 50 per cent mortality, even when carefully managed. The patients are normally elderly and frail, often have underlying conditions, particularly sepsis, and are at high risk of developing cerebral oedema, acute renal failure, and hypostatic pneumonia. The patient has normally taken some weeks of illness to reach the hyperosmolar state, and it usually requires some days at least to return them to normal.

The diagnosis may not be suspected clinically, but becomes evident when blood glucose and serum sodium levels are available in the work-up of an obviously ill patient. Most portable glucometers measure 'High' for any blood glucose >20 mmol/l.

Immediate assessment and management

Airway, Breathing, Circulation Give high-flow oxygen by facemask, insert an IV line, take blood for fbc, U&E, glucose, and blood cultures.

Begin fluid resuscitation If the patient is in shock, then begin resuscitation with normal saline, remembering that this fluid is hypotonic for the patient. However, use half normal saline for ongoing fluid management until serum Na^+ < 150 mmol/l. The fluid deficit will need to be replaced cautiously in these elderly patients—aim to give 50 per cent of the deficit in the first 12 hours and the rest in the following 24 hours.

Begin insulin therapy Seek expert advice with this condition, but generally use about half the dose used in DKA (e.g. 3 units/h).

Anticoagulate There is a real risk of deep vein thrombosis and PE. If not contraindicated, give 5000 IU heparin IV and begin infusion or arrange repeated doses.

Search for an underlying cause Perform a full history and physical examination and obtain chest X-ray, MSU, and ECG on all patients.

Disposition All such patients should be admitted to hospital under a medical or specialist endocrine team, and all should be considered for ITU care.

Euglycaemic ketoacidosis

A small proportion of young type I diabetics on insulin therapy may develop ketoacidosis without high blood glucose. Such patients may even appear relatively well, despite profound acidosis, because they do not become dehydrated. Do not mis-diagnose such a patient as having hyperventilation syndrome (see Chapter 4).

The treatment is similar to that of DKA, but the fluid requirements are much less, and dextrose/saline needs to be given early rather than normal saline.

Diabetic lactic acidosis

This is a rare complication of diabetes, usually seen in Type II diabetics. The exact mechanism of excess production and inadequate breakdown of lactate is not understood. Management is supportive with treatment of any precipitant, correction of dehydration, insulin therapy, and close observation. As with DKA, severely acidotic patients (pH < 7.20) need to be considered for the ITU.

The diabetic patient with an additional emergency

Diabetes is common and carries both medical and surgical complications. Hence, it often coexists with other emergency presentations, particularly cardiovascular and cerebrovascular disease, but also a variety of infections. In turn, these may precipitate DKA. Any diabetic requiring treatment for another condition must have a thorough assessment of their diabetic status in the A&E department. At a minimum, this includes history, examination, bedside blood glucose, and urinalysis. In the presence of hyperglycaemia, urinary ketones, or systemic illness the full investigations described in DKA (p. 100) must be undertaken. Diabetic patients with a myocardial infarction or cerebrovascular accident are admitted under medical units, but pay particular attention to ensure that a thorough assessment and ongoing review occur when a diabetic is admitted under a surgical unit (e.g. for drainage of superficial abscesses). Many hospitals have a protocol to ensure such review occurs.

Addisonian crisis

In Addison's disease there is destruction of the adrenal cortex, with resultant deficiency of both mineralocorticoids and glucocorticoid production. More commonly, Addisonian crises occur in the setting of a patient who has been on prolonged steroid therapy, which produces atrophy of the adrenal cortex secondary to decreased ACTH. In such patients, acute withdrawal of steroids or an inter-current illness without increase in steroid dose leads to glucocorticoid deficiency, with less marked mineralocorticoid effects.

Addisonian crisis is characterized by lethargy, weakness, nausea, vomiting, abdominal pain, dehydration, hypotension, hyponatraemia, and hyperkalaemia. In the absence of mineralocorticoid deficiency, the fluid and electrolyte changes are less marked and it can be particularly difficult to diagnose. If the adrenals are atrophied rather than destroyed, the patient will not show the pigmentation described in Addison's disease. If the Addisonian crisis is part of panhypopituitarism, then the patient will be pale with signs of other endocrinopathies. Suspect Addisonian crisis in any shocked patient without apparent cause, and any shocked patient who does not respond to initial therapy.

Immediate assessment and management
1. Airway, Breathing, Circulation.
2. Insert an IV line and withdraw blood for: full blood count, urea, creatinine, and electrolytes, blood cultures if febrile, blood sugar, ACTH, and serum cortisol.
3. Begin fluid resuscitation with a colloid such as gelofusin.
4. Seek and treat associated illness:
– *Hypoglycaemia* Check bedside glucose test and treat if necessary.
– *Sepsis* If there is any evidence of infection, then 'blind' antibiotic treatment is appropriate.
5. Definitive investigation if the diagnosis is uncertain. In an ill patient do not withhold steroids, but give 10 mg dexam-

ethasone IV as initial steroid therapy (this does not interfere with cortisol assay). If the patient is stable, without hypotension, after discussion with senior staff, it may be appropriate to withhold steroid therapy until after a short tetracosactrin test. Give 250 µg of tetracosactrin and draw blood at 30 minutes and 1 hour for further cortisol levels.

6. Definitive steroid therapy. If the diagnosis is known, or otherwise after a tetracosactrin test, begin steroid therapy. The high doses used in other areas of medicine are not needed for long: physiological replacement rather than anti-inflammatory effects are required. Give 250 mg hydrocortisone IV and arrange for 100 mg twice daily subsequently.

7. Search for an underlying cause. Perform thorough history and examination seeking evidence of adrenal atrophy, other endocrinopathies, and inter-current (precipitating) illness.

Disposition Admit the patient to hospital under the care of a medical team, or a specialist endocrine team if available. Consider referring the patient to the ITU if there are severe electrolyte abnormalities, ongoing hypotension, or severe underlying illness.

Thyroid emergencies

Thyrotoxic crisis

Thyrotoxic crisis may occur in the setting of known thyrotoxicosis, particularly with non-compliance, or rarely may be the presentation of new thyroid disease. Excess thyroid hormone causes fever, tachycardia (often atrial fibrillation), flushing, increased metabolic rate, diarrhoea, vomiting, marked anxiety, and tremor. Elderly patients, in particular, are at risk from the cardiovascular effects.

Control of thyroid hormone secretion takes time, and in the emergency setting, the priority must be to control the cardiovascular manifestations. This is a condition that requires expert help to manage.

Immediate assessment and management
1. Airway, Breathing, Circulation.
2. Insert IV line, take blood for fbc, U&Es, blood cultures, if indicated, and thyroid function tests.

3. Give IV propranolol 1 mg slowly. Repeat up to a maximum of 10 mg total. Titrate to reduce pulse rate to reduce pulse rate to < 120.

4. Take expert advice on the use of carbimazole (usually 15–25 mg orally 4 times daily), Lugol's iodine (usually 0.5 ml orally thrice daily starting 1–2 hours later), and steroids (e.g. dexamethasone 4 mg 4 times daily).

5. Seek an underlying cause. History and examination should identify whether new disease, compliance, or intercurrent illness is the problem. ECG and chest X-ray are mandatory. Treat infection aggressively if suspected.

6. Continue supportive care with IV fluids, paracetamol, cooling fans, and sedation if necessary.

Disposition Severe thyrotoxic crisis ('thyroid storm') is life-threatening. Discuss the patient with experts in your hospital and consider admission to the ITU on the basis of cardiovascular and neurological status.

Myxoedema coma

It is rare for hypothyroid patients to present with true coma. However, patients with hypothyroidism tend to be lethargic and at high risk of non-compliance with therapy, and elderly patients are at greatest risk of metabolic effects. When myxoedema coma does occur, it may be rapid and preceded by seizures. Severe hypothyroidism is a medical emergency, with a generalized hypometabolic state, including hyponatremia, hypotension, hypothermia, and bradycardia. The diagnosis can be made relatively easily in patients with a known history of hypothyroidism or hyperthyroidism treated with surgery or radioiodine, but may be difficult if the case is a first presentation. Be alert for signs or symptoms of hypothyroidism in comatose (see chapter 6) or hypothermic (chapter 10) patients.

This is a disease with a high mortality so enlist expert help: do not attempt to manage it on your own.

Immediate assessment and management

Airway May be at risk, especially if the patient is comatose.

Breathing Give high-flow oxygen. These patients usually hypoventilate. If hypoventilation is severe, tracheal intubation may be required, maintain breathing with bag-valve-facemask, and call the ITU or anaesthetists for assistance.

Circulation Insert an IV line, take blood for glucose, fbc, U&E, blood cultures, and thyroid function tests. Take arterial blood gases seeking hypoxia, hypercarbia, and lactic acidosis. Begin fluid resuscitation with normal saline *gently* to reduce risk of left ventricular failure.

Treat hypoglycaemia, if necessary, with 50 ml of 50 per cent glucose IV.

Perform thorough history and examination, seeking particularly infections, especially urinary tract infections and pneumonia. Obtain chest X-ray, MSU, ECG, and core temperature.

Begin thyroid hormone replacement Controversy exists over whether high or low-dose thyroxine therapy is appropriate, but all doses used are much less than those used in routine replacement therapy. Seek advice from your endocrinologist or the ITU, generally beginning with 50 µg T4 (less with ischaemic heart disease) or 10 µg T3

Begin steroids, generally hydrocortisone 100 mg IV thrice daily.

Continue supportive measures, particularly warming if hypothermic.

Disposition Patients with any impairment of consciousness from myxoedema should generally be treated in the ITU. Refer for assessment early. Thyroxine works only slowly and any immediate improvement in the patient's condition will be due to supportive care. Resuscitation, treatment of hypothermia, and treatment of infection will make a difference in the A&E department.

Further reading

1. Oh, T. E. (ed.) (1990). *Intensive care manual*. Butterworths, Sydney.
2. Ho Saunders (1992). *Current emergency diagnosis and treatment*. Lange Medical Books.

Metabolic emergencies

Hyperkalaemia

Prompt management of hyperkalaemia is essential because if left untreated cardiac arrest may occur. Muscle weakness and parasthaesiae may be the only symptoms prior to cardiovascular collapse so serum potassium must be checked in all patients at risk of hyperkalaemia. These include patients with known renal impairment with an acute illness or new symptoms, those on drugs, such as ACE inhibitors or 'potassium sparing' diuretics, that may elevate the serum potassium, and patients at risk of acute renal failure or rhabdomyolysis.

Investigation

1. Urea and electrolytes. Hyperkalaemia is defined as a serum potassium >5 mmol/l Hyperkalaemia associated with symptoms or ECG changes or levels over 6 mmol/l require treatment. Some dialysis-dependent patients chronically have levels over 6 mmol/l but if in doubt, treat and contact the renal physicians.

2. ECG. Hyperkalaemia may cause a wide variety of ECG changes including; peaked T waves, flattening or loss of the P wave, PR prolongation, widening of QRS complex, nodal rhythm, and terminally a sine wave ECG slowing to asystole. It is important to realize that these changes are not necessarily sequential. Cardiac arrest may occur soon after a normal ECG so there is no room for complacency.

3. If you think rhabdomyolysis could be the cause of hyper-kalaemia, check the creatine kinase urgently, and after consultation with the local renal physicians, treat as for rhabdomyolysis (see p. 117–19)

4. Blood glucose. Ill patients may be hypoglycaemic and diabetic ketoacidosis causes an elevated K^+.

5. Blood gases. In critically ill patients, particularly those with renal impairment it is important to monitor the degree of acidosis.

6. Hyperkalaemia can occur in an Addisonian crisis. If you think this is a possibility take blood for cortisol levels and then give hydrocortisone IV (see p. 105–6).

Management of hyperkalaemia

1. Think ABC. Has the patient arrested? Are they unconscious?

2. Give high-flow oxygen and obtain secure IV access, move the patient to a high-dependency area, and connect to a cardiac monitor. Get expert help early.

3. If the patient is in extremis with an arrhythmia or grossly abnormal ECG give calcium chloride 10 mls 10 per cent IV stat. This has no effect on the serum potassium but reduces its cardiac toxicity. This may be repeated on two occasions if necessary.

4. Dextrose and insulin. 50 mls 50 per cent dextrose with 10 units soluble insulin IV. This causes the potassium to shift into the cells, lowering the serum potassium, but does not reduce the body's total potassium load.

5. Sodium bicarbonate. 50 ml 8.4 per cent. This also causes the potassium to move into the cells and is particularly effective in patients with a coexistant metabolic acidosis. It contains a large amount of sodium and therefore must not be used in patients with pulmonary oedema as it can cause a further deterioration in lung function.

6. Haemodialysis. When there is hyperkalaemia due to acute or chronic renal failure haemodialysis is the definitive treatment so involve the renal physicians early. If the patient will have to be transferred to another hospital for haemodialysis ensure that the transfer is safe with senior medical staff as escort (see p. 193). For patients who have not previously been on dialysis, take blood to urgently check their hepatitis B and C and HIV status, and send the results to the receiving hospital.

7. Calcium resonium can be given orally 15 g thrice daily or rectally 30 g once daily, to bind potassium in the gut in patients with less severe hyperkalaemia. It is contraindicated in hypercalcaemia and should only be used by those familiar with its use.

After giving any potassium-lowering treatment recheck the serum potassium to assess response.

Metabolic acidosis

Metabolic acidosis occurs when there is an excess production of non-carbonic acid or abnormal loss of bicarbonate. Blood gas results show H$^+$ high, base excess negative, $P\text{CO}_2$ low, bicarbonate low.

Box 8.1 Causes of metabolic acidosis

- Type A lactic acidosis (due to shock and poor perfusion)
- Diabetic ketoacidosis
- Type B lactic acidosis (precipitated by ethanol, liver disease, or rarely metformin)
- Uraemia, acute or chronic
- Aspirin overdose, intentional or accidental
- Ingestion of methanol or ethylene glycol
- Renal tubular acidosis

Investigation of metabolic acidosis

Several causes of metabolic acidosis may coexist (e.g. diabetic ketoacidosis with renal impairment and sepsis) so it is essential to perform some further investigations.

Blood glucose Do not rely on glucose sticks alone as they can be inaccurate.

Urea and electrolytes Uraemia must be excluded as a cause of metabolic acidosis and hypoperfusion may result in renal impairment.

Salicylate level if there is no other obvious cause for the metabolic acidosis.

A good history and a high index of suspicion Methanol and ethylene glycol levels can be measured but these tests are unlikely to be rapidly available so consider the diagnosis particularly in alcoholics or teenagers.

Lactate It may be possible to measure the serum lactate to monitor the severity and response to treatment of lactic acidosis. It is not an essential investigation.

Amylase Pancreatitis can cause lactic acidosis.

Treatment of metabolic acidosis

Patients with metabolic acidosis have impaired cellular function and may have an impaired conscious level so after assessing and treating ABC all patients require high-flow oxygen, secure IV access, and monitoring in a high-dependency unit. The specific treatment depends on the cause of the acidosis. See the appropriate sections of this book for management of diabetic ketoacidosis (p. 100), salicylate toxicity (p. 137), and uraemia (p. 114).

Lactic acidosis due to inadequate cellular oxygenation (type A)

This is the most common cause of metabolic acidosis in the A&E department. It occurs in patients with inadequate cellular oxygenation and may result from hypovolaemic, cardiogenic, septic or anaphylactic shock, seizures, cocaine toxicity, or hyperpyrexia. When cellular oxygen delivery is inadequate anaerobic metabolism occurs with the production of excessive lactic acid that overwhelms the body's capacity to convert it to carbon dioxide.

The treatment of type A lactic acidosis is to restore tissue perfusion and oxygenation by treating the underlying condition. The airway must be clear and protected and breathing must be adequate, all patients must have high concentration oxygen. If despite high concentration oxygen the patient cannot maintain a normal Po_2 or the Pco_2 rises, the patient will require ventilation. Two wide-bore IV lines should be inserted and in all but pure cardiogenic shock volume expansion with colloid should be started to restore adequate perfusion. A urethral catheter should be inserted to monitor renal function and give a measure of the adequacy of resuscitation. Once resuscitation is started through peripheral IV lines the patient should be referred to a senior member of the in-patient medical team or ITU, as these patients are critically ill and will require invasive monitoring and probably inotropic support to optimize oxygen delivery.

Type B lactic acidosis

This used to be common in patients treated with phenformin but is unusual in patients taking metformin. The most common causes now are liver disease and ingestion of ethanol, methanol, or ethylene glycol. It is diagnosed in patients without evidence of hypoxia or hypoperfusion in whom the other causes of metabolic acidosis have been ruled out. The serum lactate is elevated.

Assessment and treatment of ABC, high-flow oxygen, and IV access are the first priorities in these patients. The cause of the lactic acidosis should be found and treated. There is no evidence of benefit from aggressive correction of pH with sodium bicarbonate. This is thought to be because acidosis actually improves tissue oxygen delivery by shifting the oxygen dissociation curve and rapid sodium bicarbonate infusion increases the affinity of haemoglobin for oxygen, thereby reducing tissue oxygenation.

Methanol toxicity (see also Chapter 9)

Methanol is metabolized to formaldehyde and formic acid that cause blindness and acidosis. After assessing and treating ABC, specific treatment includes ethanol infusion, which competitively inhibits methanol metabolism and haemodialysis. Contact the poisons information centre for advice.

Ethylene glycol toxicity (see also Chapter 9)

Ethylene glycol causes signs of intoxication similar to ethanol. Calcium oxylate crystals in the urine may cause an opalescent appearance. Oxalic acid causes severe metabolic acidosis and renal failure. Initial treatment is the same as for methanol toxicity (i.e. ABC, oxygen, and ethanol infusion). Haemodialysis will be required for patients with severe acidosis or renal impairment. Contact the poisons' information centre for advice.

Acute renal failure

The causes of acute renal failure (ARF) can be divided into three groups.

1. **Pre-renal**—due to poor renal perfusion.
2. **Renal**—due to renal disease or renal damage from toxins.
3. **Post-renal**—due to obstruction (anywhere from the renal pelvis to the urethral meatus).

Some causes of ARF, particularly pre-renal and post-renal, may be readily treatable, preventing the development of irreversible renal damage.

Management

Assess and treat ABC Give high-flow oxygen and obtain IV access. Avoid IV cannulation or venepuncture in either forearm to preserve veins that may be required for a fistula. The dorsum of the dominant hand is the preferred site.

History A full history is essential. Is there a history of diabetes, prostatism, haemoptysis, recent illness? Is there anything to suggest rhabdomyolysis (see p. 117)? Drug history of prescribed, over-the-counter, illegal, and traditional drugs.

Examination Full examination including rectal and vaginal examination. Specifically look for signs of fluid depletion (jugular venous pressure, skin turgor, mucous membranes) or overload (leg or sacral oedema, crackles, gallop rhythm), arteritis, hypertension, and compartment syndrome.

Urethral catheter This will relieve urethral obstruction and allow urine output measurement. Remember that the urine output may be normal in ARF.

Central venous pressure (CVP) monitoring Hypovolaemia is a common cause of ARF which must be corrected to prevent acute tubular necrosis. Volume replacement is usually performed with CVP monitoring. Patients with acute renal failure tolerate the complications of central line insertion less well than most patients and the technique is more difficult in hypovolaemic patients. If it is felt to be necessary, central line insertion should be performed by the most experienced staff available.

Prescribing Before giving any drug consult the *British*

National Formulary to ensure it is not nephrotoxic and that the dose does not need to be adjusted. In particular never give NSAIDS or tetracycline and only give aminoglycosides or diuretics on the advice of the renal unit.

Investigation

Urgent blood tests fbc, U&E, glucose, creatinine, blood cultures, hepatitis B markers, HIV, hepatitis C (pre-dialysis), clotting screen, osmolality, creatinine kinase (if you suspect rhabdomyolysis).

Urine Urinalysis, urgent microscopy for casts and organisms, culture, and sensitivity, osmolality.

Imaging Chest X-ray, renal ultrasound to rule out an obstructive cause (this may be done after transfer to the renal unit).

Non-urgent blood tests Autoimmune screen, antistreptolysin (ASO) titres,

Management of complications

In the presence of any of the following complications refer the patient urgently to the renal unit.

Fluid overload Usually presents as pulmonary oedema. Management is supportive, oxygen, venodilation to reduce preload if the blood pressure is adequate, and urgent referral to the renal physicians for dialysis and ultrafiltration.

Hyperkalaemia Always consider whether rhabdomyolysis could be causing the renal failure. Refer urgently to the renal unit for dialysis and see p. 109 for management while awaiting transfer.

Pericarditis Do not give non-steroidal anti-inflammatory drugs for analgesia as they are nephrotoxic. Refer urgently to the renal unit for dialysis.

Who needs referral to the renal unit?

All renal units would prefer to have patients discussed with them early rather than waiting until the patient is in extremis, in urgent need of dialysis. If in doubt, after discussion with your hospital's medical team, phone up the renal unit for advice. Any patient with ARF without an easily reversible cause, with fluid overload, pericarditis, hyperkalaemia, or a creatinine over 400 µmol/l must be referred urgently.

Chronic renal failure

It is essential that great care is taken of any existing fistula or a site where a fistula could be created, even in patients who have received a transplant. Never take blood or insert a cannula near a fistula or in either forearm. Ideally, use the back of the hand or the antecubital fossa of the dominant arm and do not apply a compressive bandage or plaster of Paris over a functioning fistula. When prescribing for a patient with renal impairment always check the formulary for potential problems, as drugs that undergo renal excretion may accumulate and cause severe toxicity. If in doubt discuss the patient with the renal unit.

Fluid overload Patients with renal impairment may present with severe pulmonary oedema. The definitive treatment is ultrafiltration, so urgent referral to the renal unit is required, but if the patient is in extremis, ventilation, or venodilation with nitrates may be indicated, after discussion with the renal unit.

Pericarditis may occur, presenting with positional, pleuritic chest pain. Always consider other diagnoses, such as ischaemic chest pain or pulmonary embolus. The treatment for pericarditis due to chronic renal failure is dialysis, so refer urgently to the renal unit. Pericardial effusion may occur which is treated with dialysis and ultrafiltration. If there is cardiac tamponade pericardiocentesis may be required, but unless the patient has had a cardiac arrest, this should be done after discussion with the renal unit.

Peritonitis due to infection of the dialysis fluid, is a common problem in patients on continuous ambulatory peritoneal dialysis (CAPD). It presents with abdominal pain and 'cloudy bags' (the effluent is usually totally clear). It is essential that this is managed promptly with appropriate intraperitoneal antibiotics to preserve the peritoneum. All patients should be urgently referred to their renal unit.

Sepsis is common in patients with CRF. Refer to the renal unit as you start resuscitation with oxygen and airway management, obtain IV access and arrange an urgent portable chest X-ray. Start antibiotics after discussion with the renal unit.

Hyperkalaemia (see p. 109).

Rhabdomyolysis

Rhabdomyolysis occurs when muscle injury results in release of its contents into the circulation. It is an emergency requiring complex management. Refer to the renal physicians early.

Box 8.2 Common causes of rhabdomyolysis

- Trauma: 'Crush syndrome' following limb entrapment, major trauma with muscle injury
- Unconsciousness: following a CVA or an alcoholic binge or overdose
- Prolonged muscular activity e.g. marathon runners and military recruits, fits
- Drugs: cocaine, ecstasy, amphetamines, often associated with hyperpyrexia
- Ischaemic muscle: following arterial emboli or surgery or compartment syndrome
- Hyperpyrexia

Rhabdomyolysis is often multifactorial, such as at 'raves', where there is prolonged vigorous exercise in a hot environment with little to drink and drugs, such as Ecstasy, are used.

Complications of rhabdomyolysis

- Acute renal failure
- Metabolic abnormalities (hyperkalaemia, hyperphosphataemia, hypocalcaemia, hyperuricaemia)
- Compartment syndrome
- Disseminated intravascular coagulation.

Diagnosis

Take a full history, including a drug history. The history from ambulance personnel or relatives may suggest a period of unconsciousness. Enquire about muscle pain or weakness (present in 50 per cent). Examination is usually normal but there may be muscle swelling or tenderness. A high index of suspicion is essential.

Investigation

Urine dipstick If myoglobin is present in the urine there will be a positive dipstick test for blood but as there must be a large amount of myoglobin in the blood for myoglobinuria to occur, up to a quarter of patients with rhabdomyolysis have normal urine dipstick results. Note the urine pH. With heavy myoglobinuria the urine may look 'smoky'.

Blood tests

Creatinine kinase. CK is the definitive test for rhabdomyolysis. Levels of five times normal are diagnostic of rhabdomyolysis, and rising levels indicate ongoing muscle damage.

Urea, electrolytes, and creatinine. It is essential to know the potassium level and to monitor renal function.

Calcium, phosphate, urate Hypocalcaemia, hyperphosphataemia, and hyperuricaemia may occur.

Full blood count, clotting screen, and fibrin degradation products DIC frequently occurs with severe rhabdomyolysis.

Arterial blood gases Patients with rhabdomyolysis are critically ill and may have a severe metabolic acidosis.

ECG and chest X-ray should be ordered. If there is any suspicion that ARF may be due or partly due to another cause the renal physicians will want to perform an renal ultrasound to rule out obstruction.

Treatment

Resuscitation Rhabdomyolysis often occurs in patients who are critically ill or injured. It is essential to evaluate and treat Airway, Breathing, and Circulation. All ill patients require high-flow oxygen and secure IV access. When obtaining IV access and taking blood avoid the forearm veins as some of these patients will develop chronic renal impairment.

Fluids Myoglobin is not directly nephrotoxic but in the presence of dehydration and acidosis a subunit of myoglobin results in ARF. Large volumes of normal saline should be given as soon as possible. In trapped trauma patients this should be started during efforts to free them. Catheterize the patient to measure urine output and aim to achieve a urine output of 200–300 ml per hour. Up to 20 litres may be required in the first 24 hours and this requires close haemodynamic and electrolyte monitoring. Start treatment with IV normal saline as soon as possible and refer the patient urgently to the renal physicians. All patients with rhabdomyolysis require admission to an ITU or high-dependency unit in a hospital with a renal unit.

Sodium bicarbonate Myoglobin is nephrotoxic when the urine is acid. Sodium bicarbonate is given to alkalinize the urine. Having checked urine pH, electrolytes, and blood gases give 50 ml 8.4 per cent sodium bicarbonate (50 mmol) and discuss the patient with the renal physicians. Patients may require over 600 mmol sodium bicarbonate in the first 24 hours.

Hyperkalaemia (see p. 109)   Management of the rhabdomyolysis with fluids and sodium bicarbonate will lower the potassium. If urgent treatment is required give calcium chloride 10 ml 10 per cent IV for cardiovascular collapse. Otherwise discuss with the renal physicians.

Surgery If there is a suspicion of compartment syndrome refer to the orthopaedic team for compartment pressure monitoring. If this confirms elevated compartment pressure fasciotomy must be performed. If there is extensive dead muscle leading to rhabdomyolysis and hyperkalaemia this must be debrided if the patient is to survive.

Dialysis If patients present late in established renal failure dialysis may be required.

Hypercalcaemia

The most common causes of severe hypercalcaemia are hyperparathyroidism and malignancy. It may also occur in Addison's disease, sarcoid, and during treatment with thiazides. Untreated, it may lead to coma, cardiovascular collapse, and death.

Clinical features

Symptoms usually develop with total serum calcium levels above 3.5 mmol/l. Muscle weakness, tiredness, and itching; mental disturbances, including confusion, anxiety, or psychosis; and gastrointestinal disturbances, particularly vomiting, may occur. Polyuria, polydipsia, and thirst result from impaired renal water resorption.

Investigation

Urea and electrolytes, creatinine, phosphate, magnesium, fbc, chest X-ray, and ECG should be checked urgently. Blood should also be taken for alkaline phosphatase, albumin, LFTs, and plasma protein electrophoresis.

Management

Emergency management consists of rehydration. Calcium-lowering drugs and specific treatment of the underlying condition can be used later.

Rehydration This is the most important part of the management of hypercalcaemia: 4–6 litres of normal saline given over 24 hours will correct dehydration and inhibit calcium resorption. Central venous pressure monitoring enables fluid overload to be avoided. Once rehydration is achieved frusemide is given with the fluid to prevent fluid overload and increase calcium excretion. Potassium and magnesium must be closely monitored as supplements are likely to be required.

Calcium-lowering drugs Steroids (e.g. prednisolone 40–60 mg daily) and subcutaneous calcitonin (e.g. salcatonin 100–300 units thrice daily) can be effective temporarily.

Biphosphanates are effective in hypercalcaemia due to malignancy. They start to have an effect within 48 hours, normocalcaemia being achieved within a week. The choice of drug should be made by the in-patient team.

Treatment of the underlying cause (e.g. surgery, radiotherapy) It is important that treat is not withdrawn from patients with malignancy on the basis of their poor quality of life while they remain hypercalcaemic. Physical and psychological symptoms may disappear once the calcium is normal.

Further reading

1. Oh, T. E. (1990). *Intensive care manual.* Butterworths, Sydney.
2. Janson, C. L. and Marx, J. A. (1992). Fluid and electrolyte balance. In *Emergency medicine: concepts and clinical practice* (ed. P. Rosen). Mosby-Year Books, St. Louis, MO.

CHAPTER 9

Poisoning

- **Introduction 121**   **Drugs of abuse 126**   **Other drugs 132**
  Chemicals 141

Introduction

Initial assessment

When a patient presents having taken a substance you know nothing about, it is tempting to rush away and look it up. Before you do, **stop and think ABC.** Use universal precautions including gloves, and, where appropriate, eye protection, at all times.

The airway   Every poisoned patient needs a patent, protected airway. Use simple manoeuvres to open the airway (see Chapter 2) If you think the airway needs to be protected by tracheal intubation, call an anaesthetist. Intubation in this setting is best left to the experts if possible, as the stomach is usually full and vomiting is common.

Breathing   Many drugs reduce respiratory drive. If the patient is not breathing open the airway and start ventilation. Give every patient with an impaired conscious level high-flow oxygen, and check the arterial blood gases. Patients who are breathing, but with inadequate respiratory effort may need intubation and ventilation.

Circulation   Check for a pulse. If the patient has had a cardiac arrest start cardiopulmonary resuscitation. Any patient with an impaired conscious level should have an IV line inserted. Check the heart rate and blood pressure, if abnormal arrange for a 12-lead ECG. Check BM stix for capillary blood sugar.

Get a brief history What has the patient had taken? When? Any empty bottles in the room? What drugs are kept in the house? If necessary, get the family or police to search the house.

What next? There are hundreds of thousands of drugs, chemicals, and household products. No one can know the up-to-date management of all of them and this is why poisons' information centres exist. The telephone numbers of all the UK centres are in the *British National Formulary.*

You may know nothing about the drug or chemical your patient has taken but if you remember **ABC** you can keep them alive while you find out.

Toxicological investigation Unfortunately, although many drug levels can be measured there is no rapidly available 'drug screen' that can be used in patients who have taken an unknown substance. Physical examination and a clear drug history of the both the patient and others in their household may suggest likely toxins so that specific drug levels can be requested.

Useful investigations in all ill patients following poisoning include bedside and formal blood glucose, urea and electrolytes, and arterial blood gases. Paracetamol is readily available, commonly taken in overdose and treatment must be instituted early if it is to be effective. For these reasons a paracetamol level is usually checked in patients who have taken an unknown overdose.

Other drug levels that may be of benefit are those where a rapid test is available and treatment depends on the drug level. Examples include, theophylline, digoxin, iron, lithium, and aspirin. If in doubt discuss the patient with the poisons' information centre.

Removing the toxin

When a patient presents having taken a potentially toxic overdose emptying the stomach of a toxin may seem at first sight a sensible first intervention, but always remember that ABC must take priority.

If you are uncertain which would be the most effective method of removing the toxin from your patient, ask the poisons' information centre for advice.

Activated charcoal   Activated charcoal (50 g) is administered by mouth of through a naso-or orogastric tube as a slurry. In adults following poisoning it is replacing emetics and gastric lavage as the preferred method of removing toxin from the body, although, because children do not willingly drink it, it is used less in paediatric practice. Activated charcoal acts as an adsorbent for many drugs and toxins, absorbing both from the stomach and small bowel lumen and also from the circulation across the small bowel mucosa, reducing their half-life.

Box 9.1 **Drugs *not* absorbed by activated charcoal**

- Simple ions (e.g. heavy metals, including iron, lithium, and cyanide)
- Strong acids and bases
- Alcohols (e.g. ethanol and methanol)
- Hydrocarbons

There are some drugs which are absorbed so effectively that repeated doses of activated charcoal, 50 g 4 hourly are recommended. These include: theophylline, barbiturates, digoxin, aspirin, carbamazepine, tricyclics, and quinine. It must be remembered that as in all patients with poisoning, when giving activated charcoal the airway must not be only open but protected. Aspiration of charcoal in unconscious patients can kill.

Gastric lavage   For most drugs, there is no point in emptying the stomach more than 4 hours after ingestion as little drug will remain in the stomach. Important exceptions to this are aspirin and tricyclic antidepressants, which delay gastric emptying, and some slow-release formulations, particularly theophyllines. If in any doubt discuss the patient with the poisons' information centre.

Gastric lavage has been the traditional way of emptying the stomach following poisoning. With the patient in a head-down left lateral position a large bore lavage tube is passed through the mouth into the stomach. The stomach is then washed out with water until no tablet debris can be seen. If the drug will be absorbed by activated charcoal (see above), activated charcoal (50 g) should be put down the lavage tube and left in the stomach, before the tube is removed.

If a patient has an impaired conscious level or gag reflex gastric lavage can only be done with a cuffed tracheal tube in place, supervised by an anaesthetist. It may be necessary to anaesthetize a semi-conscious patient to allow the procedure to be carried out safely.

Gastric lavage is unpleasant for a conscious patient and hazardous for a patient with impaired conscious level if the airway is not protected, so it should be reserved for patients who have taken a serious overdose. It must never be used as a punitive measure to discourage self-poisoning. Gastric lavage is contra-indicated in poisoning with both caustic substances when it may worsen oesophageal burns, and oil-based substances when even minor aspiration can lead to chemical pneumonitis.

Ipecacuanha (ipecac)   This is a drug that is extremely effective at inducing vomiting, but it should not be used over 4 hours following drug ingestion, as by this time little drug will remain in the stomach (see above). For those drugs which are adsorbed by activated charcoal, ipecac is thought to be less effective at reducing the body's drug load than activated charcoal, and as the two cannot be used together, activated charcoal should be given in preference to ipecac.

Box 9.2 Contraindications to ipecac

- Caustic or oil-based substance taken in overdose
- Altered conscious level
- Substance taken could alter conscious level (the patient may vomit and aspirate)
- Substance taken would be absorbed by activated charcoal

Assessing mental state after deliberate self-harm

Resuscitation takes priority in patients following deliberate self-harm, but it is important that management is not limited to the physical problems.

Psychiatric assessment in A&E   In severely poisoned patients, the psychiatric assessment will occur on the ward once the patient has recovered. In some A&E departments, psychiatric assessment is performed by senior A&E staff, in patients who have taken a non-harmful overdose. This can be acceptable practice if certain conditions are met:

- Patient's behaviour not altered by alcohol, benzodiazepines, or other drugs.
- Doctor trained in assessing patients following deliberate self-harm.
- The patient must feel the doctor has all the time in the world for them and is interested in what they have to say.
- Privacy. The interview should be in a room, not a curtained cubicle. There should be a panic button for the doctor to summon help if required.
- Relatives, friends, and police officers should not be present.
- If the patient does not have a common language with the doctor, in which both are fluent, there must be an interpreter who is not a relative or friend of the patient.

Certain patients **must** be seen by a psychiatrist. These include, those aged under 16, patients in the puerperium, those with psychiatric illness, or who are suicidal.

Box 9.3 Suicide risk factors in deliberate self-harm

- Violent suicide attempt (e.g. jumping, near-hanging)
- Known or newly diagnosed psychiatric illness
- Recent loss (e.g. bereavement, divorce, redundancy, retirement)
- Puerperium (ensure baby is safe and being cared for)
- Age > 40, male, social isolation, alcoholism
- Premeditated episode with suicide note.
- The patient thought it would be lethal and expected to be undisturbed for some hours

---

Self-discharge Some patients who you consider to be psychiatrically ill will want to leave the hospital against medical advice. If you consider they are a danger to themselves or others, even if the psychiatrist has not arrived, **do not let them go**. Under common law you can hold them against their will until the section papers have been signed. If someone you consider sectionable and a danger to themselves or others leaves the department, inform the police and get them brought back.

Discharge from A&E If no treatment or monitoring is required for the poisoning, patients may go home if there are no symptoms or signs of psychiatric illness, no risk factors for suicide, a responsible adult at home, the patient's GP is aware, and follow-up has been arranged. It is essential that all patients should be followed-up following an episode of deliberate self-harm. Those with psychiatric illness, not felt by the psychiatrist to require in-patient care should have psychiatric follow-up arranged before they leave the hospital. Those patients who are not psychiatrically ill or suicidal should have general practitioner follow-up arranged.

Poisoning: drugs of abuse

Alcohol

In A&E departments it is very common to see patients who smell of alcohol and are confused or unconscious. Admittedly, alcohol can make you confused or unconscious but so can other conditions that require urgent treatment. Alcohol intoxication is a diagnosis of exclusion.

Assess and treat ABC Even if the patient is 'only drunk', airway obstruction kills. Continue to reassess ABC as the patient may deteriorate after presentation. If the patient is unconscious give high-flow oxygen.

Check the blood sugar Hypoglycaemia can mimic drunkenness and acute alcohol toxicity may cause hypoglycaemia, particularly in the young. Hypoglycaemia associated with alcohol intoxication should be treated with IV dextrose, not glucagon. In patients with chronic alcohol problems also give thiamine 100 mg IV to prevent Wernicke's encephalopathy.

The alcohol level does not matter It is possible to obtain an accurate measurement of the blood alcohol level from the laboratory, or, in patients who are able and willing to co-operate, an instant breath alcohol level, but these are of little value in patient management. Do not let the blood or breath alcohol affect your diagnosis or management. Tolerance differs so a high level does not tell you that unconsciousness is due to alcohol and if the airway is in jeopardy it needs protection regardless of the blood alcohol level. (See Box 9.4) Always do a full head-to-toe examination for signs of injury.

Box 9.4 **Alcohol intoxication—think about and exclude:**

- Head injury. This frequently occurs in intoxicated patients
- Hypoxia and other medical or neurological conditions
- Opioid or other drug toxicity: Give naloxone IV if it is a possibility (naloxone only reverses opioids)
- Ethylene glycol or methanol ingestion (see p. 143)

An intoxicated patient in the ITU? The indications for admission to the ITU do not change because the illness is due to alcohol. If the airway is not maintained without intubation, ventilation is inadequate or there is circulatory compromise refer the patient to the ITU.

The intoxicated patient with a head injury Patients have died of intracranial haematomas because their physical signs of unsteady gait, nystagmus, aggression, and impaired conscious level were assumed to be due to alcohol intoxication.

It is essential that patients with any evidence of head injury are investigated and treated as if all their symptoms and signs were due to that head injury, by medical staff trained in the management of head injured patients. The indications for a skull X-ray, CT scan, and neurosurgical referral remain identical even if the patient has been drinking.

If you are not trained in the assessment and management of head-injured patients, call for experienced help immediately and use the ABC approach: Airway with cervical spine immobilization, Breathing, and Circulation with haemorrhage control.

The ongoing management of patients with a suspected head injury is outside the remit of this book but is well described in other texts.

Opioids

Opioid toxicity may occur as a result of recreational use, self-poisoning or over-enthusiastic prescribing. It is easily treated so consider it in every unconscious patient. If in doubt give naloxone, it will do no harm.

Presentation Opioid toxicity causes loss of consciousness with airway obstruction, respiratory depression with reduction in both the respiratory rate and the tidal volume, culminating in respiratory arrest, reduced cardiac output, and eventually cardiac arrest from hypoxia. Patients may be any age and may not have a known history of opioid use. The pupils may not be small.

Management

Ensure your own safety Use universal precautions and get help early.

Airway Open the airway by a head-tilt/chin-lift or jaw-thrust.

Breathing Give high-flow oxygen. If the patient is apnoeic or breathing is inadequate, start ventilation.

Circulation Check for a central pulse. If the patient is pulseless start cardiopulmonary resuscitation. If there is a pulse obtain IV access.

Naloxone This opioid antagonist should be given IV at an initial dose of 1.2–2.4 mg. If there is an incomplete response give further doses—4 mg or more may be required. At this point the patient usually wakes up and may become agitated. Ensure you have help.

The half-life of naloxone is much less than those of most opioids so if you did nothing more the patient would become unconscious again, and possibly have a further respiratory arrest. Prevent this by a naloxone infusion: 2 mg naloxone in 500 ml 5 per cent dextrose, the rate titrated against the conscious level.

Reassess ABC If the patient has been hypoxic for some time or has taken a mixed overdose they may not wake up immediately. If their airway is not protected or breathing is inadequate they will require intubation and ventilation.

Consider mixed overdose Many patients take mixed overdoses, so look for evidence of other drugs. Some analgesic preparations (e.g. coproxamol, codydramol, cocodamol, and many others) combine opioids with paracetamol. If there is any doubt check the paracetamol level.

Continuing management Despite reversal of the opioid, patients may remain critically ill because of the preceding period of unconsciousness and/or hypoxia. Hypoxic brain injury may lead to persistent coma, requiring airway protection and ventilation. Pulmonary aspiration may have occurred while the patient was unconscious, and large doses of some opioids may cause pulmonary oedema. Patients who have had a prolonged period of unconsciousness are also at risk of rhabdomyolysis and compartment syndrome. A full head-to-toe examination is essential and if rhabdomyolysis is suspected, blood should be taken for an urgent creatinine kinase measurement (see Chapter 8).

Amphetamines and Ecstasy

The use of amphetamines and Ecstasy is increasing. Many users only experience the toxic effects they hoped for, that is elimination of fatigue and a feeling of well-being, but in a small number the effects can be life-threatening.

Sudden death may occur due to arrhythmias, so always check ABC first.

Behavioural problems Hyperactivity and hallucinations are common. Put your safety first and do not risk injury. Attempt to manage the patient by calm reassurance but if they are totally unmanageable get help to restrain them and sedate with IV diazepam. Do not use prolonged physical restraint as it is inhumane and fighting against it will worsen their fever and rhabdomyolysis.

Hyperpyrexia Not only do these drugs cause pyrexia but they are often taken in clubs where it is hot, drinks are expensive, and dancing goes on for many hours. These combine to cause dehydration and marked hyperpyrexia. Treatment is aggressive rehydration with cool IV fluid and cooling with tepid sponging and fans (see the section on heat illness, p. 148).

Rhabdomyolysis If rehydration is delayed or inadequate, rhabdomyolysis can result in acute renal failure. After starting rehydration measure the urea and electrolytes and creatinine kinase (CK) and check for myoglobinuria. If there is myo-globinuria or a markedly elevated CK seek urgent help from senior ITU or renal medicine staff. While waiting for their arrival continue treatment with supplemental oxygen and IV fluids. Take blood for ABG, urate, and arrange for an ECG. (See also p. 117 for the management of rhabdomyolysis.)

Other problems Uncommon problems include seizures, ischaemic chest pain, and cerebral haemorrhage. In all cases, assess and manage ABC first and remember to check for fever and rhabdomyolysis.

Cocaine/Crack

Cocaine and freebase cocaine (crack) use causes problems due to their sympathomimetic effects and direct neurotoxicity (see Box 9.5).

Box 9.5 Problems caused by cocaine

- **Psychiatric**: hallucinations, and anxiety, causing violent behaviour.
- **Cardiovascular**: hypertension, myocardial ischaemia/myocardial infarction, arrhythmias, aortic dissection
- **Neurological**: seizures, cerebral/subarachniod haemorrhage, transient ischaemic attack/cerebral infarction
- **Other**: hyperpyrexia, rhabdomyolysis, placental abruption

---

Management

Safety first Your own safety and that of other staff must be ensured. Always use universal precautions and never try to be a hero.

ABC Always check and treat ABC. Check the blood sugar as hypoglycaemia may cause a similar picture. Give high-flow oxygen. If you find one cocaine-related problem look for others. If a patient is unconscious give IV naloxone (see p. 128). It has no effect on cocaine toxicity but the patient could have also taken an opioid.

Seizures These **must** be controlled. Treat initially with IV diazepam, 10 mg initially, repeated if necessary. Use phenytoin if diazepam is ineffective. Referral to the ITU for thiopentone, paralysis, and ventilation may be required.

Agitation/violence These patients are dangerous. A psychotic violent cocaine-user will need many trained people to restrain them so get help quickly, from hospital security staff and the police, then chemically restrain with IV diazepam. Avoid phenothiazines as they may precipitate fits, and prolonged physical restraint as it will worsen agitation, hyperpyrexia, and rhabdomyolysis. Treat hyperpyrexia if present.

Chest pain Chest pain associated with cocaine use must be treated seriously as coronary artery spasm can cause myocardial infarction. Give high-flow oxygen, get IV access and treat agitation or hyperpyrexia. Check a 12-lead ECG and even if it is normal arrange admission to the cardiac care unit for monitoring, ECG, and cardiac enzyme series.

Hyperpyrexia Agitation or fits contribute to hyperpyrexia. Control them with IV diazepam and cool the patient with tepid sponging and a fan. Rehydrate with cold IV fluid. Look for rhabdomyolysis and treat if present (see p. 117).

Hallucinogens

Lysergic acid diethylamide (LSD) is a very potent hallucinogen. No human deaths have been attributed to its physical toxicity, but the psychological effects have resulted in death from accidents, suicide, and murder. The most common hallucinogens used are plants, such as 'magic mushrooms' (*Psylocybe* spp.) and nutmeg, which do not have the potency of LSD.

Treatment
Check ABC Rarely severe toxicity with fits or hyperpyrexia can occur.

Full assessment Are all the symptoms and signs due to a hallucinogen or does the patient have any signs of injury, an organic confusional state or signs of toxicity from other drugs?

Ensure the patient's safety The patient must be protected from accidental injury and self-harm until their symptoms subside. If the patient is distressed they should be in a quiet environment and be reassured that their symptoms are drug-induced and will resolve. On the rare occasion that a patient is uncontrollable with severe agitation, you will need help from hospital security or the police. The patient will require sedation with IV diazepam or haloperidol. Do not use physical restraint other than to give the sedation as it is inhumane, and struggling will cause acidosis and hyperpyrexia.

Other drugs

Paracetamol

After ingestion of excessive doses of paracetamol the usual route of metabolism (i.e. biotransformation by the hepatic microsomal enzymes and conjugation with glucuronic and sulphuric acid) is overloaded, and metabolism occurs through the cytochrome P-450 mixed-function oxidase system. The resultant biochemical products are directly hepatotoxic. These products are usually detoxified by conjugation with gluta-thione but in large overdoses the glutathione stores are exhausted and hepatocyte destruction occurs.

N-acetylcysteine replenishes the stores of glutathione, thus preventing hepatic damage but its efficacy starts to decline at 8 hours post-ingestion, so, when indicated, treatment with N-acetylcysteine should be started as soon as possible.

Which patients need a paracetamol level?   All patients who are known to have taken more than the recommended dose of paracetamol must have a paracetamol level checked at 4 hours post-ingestion, or immediately if they present more than 4 hours post-ingestion.

Because paracetamol poisoning is potentially fatal but easily treated, a serum paracetamol level should also be taken on any patient who you suspect may have taken an overdose. This includes unconscious patients following overdose and those in whom there is any doubt about which drugs have been taken.

Assessing the severity of poisoning   By knowing the plasma paracetamol level and the time since ingestion it is possible to assess the risk of liver damage. Blood should be taken for a paracetamol level, not less than 4 hours after ingestion, and the result plotted on the paracetamol treatment graph (see Fig. 9.1). If your patient's paracetamol level falls above the line joining 200 mg/l at 4 hours post-ingestion and 50 mg/l at 12 hours post-ingestion they are at risk of liver damage and treatment with N-acetylcysteine should be started immediately.

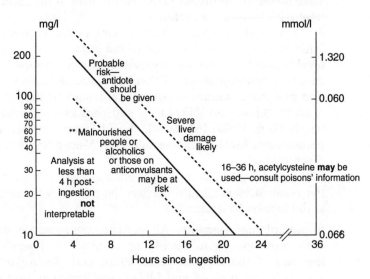

** Guide only, not based on published data.

Fig 9.1 • Paracetamol treatment graph. From the National Poisons Information Service (1993).

Some patients may be more susceptible to liver damage and should be treated with *N*-acetylcysteine when their paracetamol level is 50 per cent or more of the normal toxic level. This is represented on the paracetamol treatment graph by the lower dotted line. Patients who should be treated at these lower paracetamol levels include those with induced liver enzymes (alcoholics and patients on rifampicin or anti-epileptic medication) and those who may have glutathione deficiency (the malnourished and patients with anorexia or HIV).

If you cannot be certain at exactly what time the paracetamol was taken, or if the tablets were taken over a period of time, assume the worst and treat the patient as if the paracetamol was taken at the earliest likely time.

Management

<4 h post-ingestion   If there is a history of a significant overdose empty the stomach using gastric lavage and check the paracetamol level at 4 h.

4–8 h post-ingestion   Gastric lavage will not be of benefit.

Take blood for an urgent paracetamol level. If the level is toxic treat with IV *N*-acetylcysteine.

8–24 h post-ingestion Start *N*-acetylcysteine immediately, without waiting for the result of the paracetamol level. If the level is non-toxic the *N*-acetylcysteine can be stopped.

24–36 h post-ingestion If there is a strong history of ingestion of a toxic amount of paracetamol, start treatment with *N*-acetylcysteine. The INR, LFTs, and paracetamol level should be checked. If the INR is <2, the LFTs are normal, and the paracetamol level is undetectable the *N*-acetylcysteine can be discontinued.

>36 h post-ingestion If the patient has signs of hepatotoxicity the medical team should discuss the patient with senior staff at the nearest liver unit.

In-patient management All patients who require treatment with *N*-acetylcysteine should be admitted to hospital under the care of the medical team. Fluid and electrolye balance should be maintained and LFTs, renal function, and the INR should be monitored daily.

Box 9.6 Indications for referral to a liver unit

- >36 h post-ingestion and any evidence of hepatotoxicity
- INR >2.0 at 24 h, >4.0 at 48 h, or >6.0 at 72 post-ingestion
- Any evidence of hepatic encephalopathy
- Elevated creatinine kinase
- Hypotension despite volume replacement
- Metabolic acidosis with pH <7.3

N-acetylcysteine The dose of N-acetylcysteine, given intravenously is:

---

	150 mg/kg in 200 ml 5% dextrose over 15 minutes
then	50 mg/kg in 500 ml 5% dextrose over 4 hours
then	100 mg/kg in 100 ml 5% dextrose over 16 hours

Benzodiazepines

Benzodiazepine toxicity may result from recreational use or intentional or accidental overdose. There is a progressive depression of conscious level, leading to coma. Severe respiratory depression is uncommon but may occur.

Initial management is by the ABC approach. The airway must be clear and protected, ventilation must be adequate to maintain normal oxygenation and carbon dioxide elimination. If the airway is in jeopardy or ventilation is inadequate without support these must be corrected immediately and the patient should be referred to the ITU.

It is common for patients to take a mixed overdose of benzodiazepines with alcohol and other drugs such as tricyclic antidepressants and analgesics. Therefore, although circulatory problems are uncommon with benzodiazepines alone, a full assessment is essential. Attempt to get a clear history from the family or ambulance personnel and if there is any suspicion of paracetamol having been taken check a paracetamol level.

Flumazenil, a specific benzodiazepine antagonist is not safe, or licensed for use following overdose. The reason for this is that in patients who have epilepsy, those who have taken a mixed overdose, or chronic benzodiazepine users, flumazenil may cause prolonged fitting that is highly resistant to treatment. Flumazenil is only licensed to reverse sedation in anaesthetic practice.

Antidepressants

Tricyclic antidepressants are potentially lethal drugs, and overdose is relatively common because of the patient population for which they are prescribed. The majority of deaths from tricyclic overdose occur out of hospital, and good supportive care should ensure survival of all those who are not moribund on arrival at the A&E department. In general, the older agents (amitriptyline, imipramine) are more dangerous than the new products, but the particular drug, quantity taken and serum levels are not of great prognostic value. Treatment is supportive, aimed at the expected complications:

Cardiovascular Sinus tachycardia, SVT, VT, AV block, hypotension, cardiogenic shock. The ECG may show prolongation of the QT interval, widening of the QRS complex, and bundle branch block. The cardiac toxicity is worsened by acidosis and hypoxia, and is the usual cause of death in untreated patients.

Neurological   Lowered seizure threshold, pyramidal effects, extrapyramidal effects, drowsiness, stupor, coma. Anticholinergic effects including confusion, hallucinations and dilated pupils, dry mouth and urinary retention.

The effects vary between patients, but the common severe presentations are coma and seizures. The former requires supportive measures, the latter responds to IV benzodiazepines. Airway obstruction secondary to seizures and coma is another common cause of death.

Metabolic   Hypotension and hypoventilation cause metabolic and respiratory acidosis. These exacerbate the other toxic effects and require treatment.

Treatment   Follow the general principles of treatment for overdose described on p. 121. Gastric lavage may be useful up to 6 hours or more following overdose, because of delayed gastric emptying. In patients with an impaired conscious level gastric lavage should only be performed with the airway protected by a cuffed endotracheal tube. Do not use ipecac as the patient's conscious level could deteriorate prior to emesis occurring. All patients who have taken an overdose of antidepressants should be given activated charcoal.

Investigation   All patients who may have taken a tricyclic antidepressant overdose should have a 12-lead ECG. Those with drowsiness or any ECG abnormality should have arterial blood gases performed. Given the potential for multiple drug overdose consider serum paracetamol and blood glucose. If the conscious state is abnormal make certain of the diagnosis historically or investigate as for coma (see Chapter 6).

Management of asymptomatic patients   Those who present less than 6 hours after overdose, with no symptoms or only mild drowsiness and a normal ECG should have gastric lavage performed and activated charcoal left in the stomach. They can then be admitted to a medical ward and observed for the development of further symptoms.

Patients presenting over 6 hours after ingestion without significant symptoms and with a normal ECG are unlikely to develop life-threatening complications. They will require full psychiatric assessment (see p. 124).

Management of symptomatic patients Patients with coma, seizures, neurological signs, hypotension, acidosis, or ECG abnormalities require close observation, monitoring, and treatment in hospital, preferably in an ITU or CCU. Patients with acidosis or ECG abnormalities, other than sinus tachycardia, are likely to develop life-threatening complications. Ensure their airway is open, give high concentration oxygen by reservoir facemask, obtain secure IV access, and seek senior help. The patient should be referred to the ITU. Intensive therapy management is likely to include sodium bicarbonate to correct the acidosis, intubation, and hyperventilation. If no expert help is immediately available and the patient has a severe metabolic acidosis, providing that ventilation is adequate, give sodium bicarbonate 100 mmol IV over 15 minutes.

A patient with an isolated tachycardia in the range 100–120 beats/minute but an otherwise normal ECG does not require cardiac monitoring for this condition alone. However, it is important to check for other possible causes of the tachycardia, especially dehydration, or hypotension. If you are in any doubt discuss the patient with the poisons' informations centre.

Aspirin and salicylate

Aspirin and other salicylates poison cellular metabolism by uncoupline oxidative phosphorylation. The effects of salicylate overdose are:

- CNS effects, tinnitus, restlessness, hyperventilation (with mild respiratory alkalosis).
- Metabolic acidosis with nausea, vomiting, sweating, dehydration.
- Hypoglycaemia, hypokalaemia, prolonged bleeding time.
- Seizures, hyperpyrexia, coma, death.

Patients following salicylate overdose benefit from multiple doses of activated charcoal and gastric lavage may be effective up to 12 hours after overdose.

Patients with a history of large overdose or any symptoms, should have fbc, U&E, CK, ABG, and **urgent** salicylate level taken on arrival. If salicylate is detected arrange a follow-up level in hospital at 4 hours.

All patients with symptoms, signs, a serum salicylate over 300 mg/l at 6 hours, or a chance of later toxicity (e.g. slow-release tablets) must be admitted under a medical team. Patients with a level over 500 mg/l are considered as having severe intoxication.

If you are not familiar with treating patients following salicylate overdose seek expert help and ask the poisons' information centre for advice.

Supportive treatment should include cooling if there is fever, IV hydration, and correction of hypokalaemia and hypoglycaemia. Severe metabolic acidosis should be treated with sodium bicarbonate (100 mmol IV over 20 minutes). If there is a coagulopathy give vitamin K (10 mg IV) and consider FFP. Patients with any metabolic abnormality should be considered for ITU care.

If the patient is in coma, fits, has impaired renal function, or a serum salicylate level >900 mg/l or >700 mg/l and rising rapidly, specific treatment is charcoal haemoperfusion or haemodialysis which is performed in an ITU.

Non-steroidal inflammatory drugs

The safety profiles of most non-steroidal anti-inflammatory drugs are much better than those of aspirin or salicylate. The major side-effects are those of gastrointestinal irritation, although very high doses can cause salicylate-like effects. Mefanamic acid, taken in overdose, frequently causes status epilepticus. The management is supportive with the fits being initially treated with diazepam.

Iron

Iron and lithium are the only two prescribed medications for which activated charcoal is ineffective. Severe poisoning with iron has three phases: (1) initial corrosive gastrointestinal upset with vomiting, abdominal pain, nausea, and gastrointestinal haemorrhage if severe; (2) a period of recovery 6–18 hours after ingestion; and (3) severe cellular toxicity with hepatic failure, shock, acidoses, hypoglycaemia, and subsequent coma and death. It is vital to remember that a patient who is apparently recovering on the day after overdose is actually at high risk of complications.

Investigations

Blood tests Obtain fbc, U&E, CK, glucose, baseline coagulation studies, and urgent serum iron, and total iron-binding capacity on any patient who has taken an iron overdose.

X-ray Iron tablets form concretions which may be visible on X-ray. A plain abdominal X-ray may confirm the diagnosis of iron overdose, and/or demonstrate a large concretion in the stomach, which can be removed by gastroscopy.

Management

Iron tablets come in different strengths Check the actual amount of ferrous (Fe^{2+}) iron taken—amounts >30 mg/kg are likely to be toxic. If you are dealing with a serious iron overdose, enlist the help of your ITU early and contact the poisons' information centre for advice.

Gastric emptying Perform gastric lavage. Ensure that the airway is protected and lavage until the return is clear. If there is evidence of a concretion on abdominal X-ray then consult a gastroenterologist about the possibility of removal by endoscopy. Although controversial, many experts recommend leaving 5 g of desferrioxamine in the stomach after lavage.

Chelation therapy Symptomatic patients and those in whom the serum iron level is greater than the total iron-binding capacity (TIBC) should receive chelation therapy with desferrioxamine. Give 2 g IM then 80 mg/kg over 5 hours IV daily. This needs to be continued until the urine is no longer pink (colour of chelated complex) or iron levels have fallen. All patients with significant iron overdose must be admitted under a medical team and closely monitored.

Beta Blockers

Beta blockers competitively block adrenergic action, so their most dangerous effects are on the cardiovascular system. Beta blocker toxicity causes bradycardia, atrioventricular block, widened QRS duration, and reduced myocardial contractility, producing hypotension and shock or pulmonary oedema without tachycardia, progressing to cardiac arrest. Bronchospasm can occur in asthmatics. Beta blockers also cause hypocalcaemia (by antagonizing parathyroid hormone),

hyperkalaemia, hypoglycaemia and, if lipid-soluble, sometimes seizures.

Management
Check ABC. If the patient has had a cardiac arrest start CPR.
   Call for senior help early.
   If there is bradycardia with cardiovascular collapse give
atropine 0.5−3 mg IV, and, if ineffective, try glucagon
5−10 mg IV. For resistant bradycardia consider an isoprenaline infusion (0.05 µg/kg/min, increasing to a maximum
of 3 µg/kg/min), or external cardiac pacing. Urgently check for
glucose and electrolyte abnormalities and give repeated doses
of activated charcoal.

Digoxin

Cardiac glycoside toxicity probably occurs more frequently
through therapeutic misadventure than deliberate overdosage,
and should be suspected in any patient taking digoxin who
has a new arrhythmia or EGG abnormality. Most large centres
can measure an urgent serum digoxin level if the diagnosis is
in doubt. Digoxin acts through an increase in intracellular calcium, and therefore toxicity is aggravated by coexistent hypercalcaemia, hypokalaemia, or hypomagnesaemia. Symptoms of
toxicity are nausea, vomiting, confusion, and chromatopsia
(seeing yellow). Signs include arrhythmias of any kind, frequently ventricular tachycardia, or a rapid atrial rhythm with
variable AV block.

Management   Treatment is both supportive and specific. Give
activated charcoal in repeated doses. Treatment of arrhythmias is only required if the patient is symptomatic. Treat
bradycardia with atropine 0.5−3 mg IV, ventricular arrhythmias with lignocaine 1 mg/kg IV plus infusion (see Chapter 3),
and avoid DC cardioversion if you can. If there is hypokalaemia or hypomagnesaemia give IV supplements to maintain potassium and magnesium at high normal levels.
   Specific therapy is 'Digibind', Fab fragments of antidigoxin
antibodies. This is expensive and may be difficult to obtain
but is extremely effective. The required IV dose can be calculated from the serum digoxin [0.34 × body weight (kg) × serum
digoxin (ng/ml)]. Check the product information—depending
on the preparation many ampoules may be required in a large

overdose. If in doubt seek advice from the poisons' information centre.

Carbamazepine

Carbamazepine, like the other anti-epileptic drugs, causes CNS depression, ataxia, nystagmus, and hypotension in overdosage. Repeated doses of activated charcoal may be effective, but the patient will require admission for a day or two because of the relatively long half-life. Treat symptomatically with IV fluids for hypotension and support. If CNS depression is severe, the standard indications for intubation and ITU management apply.

Chemicals

Acids and alkalis

Strong acids and alkalis are available in industry and laboratories but the common agents ingested are domestic automatic dishwater powder or liquid (pH 10–12) and drain cleaner (liquid NaOH, pH 13+). Both acids and alkalis directly damage the oral and gastrointestinal mucosa, but alkalis tend to penetrate more deeply and cause more severe burns, with the effects depending on the agent, dilution, and pH. The condition of the oral mucosa may not be a reliable guide to that of the oesophagus. In most cases, the stomach withstands the acid, so damage is confined to the pharynx and oesophagus. The immediate problems are of swelling with upper airway obstruction, bleeding, and oesophageal perforation. The long-term effects are oesophageal scarring and obstruction. Some corrosive agents, such as phenols, also have systemic effects—check with the poisons' information centre.

The immediate first aid following acid or caustic ingestion is dilution with copious amounts of water or milk. Do not use acids to neutralize alkali or vice versa, as the heat generated may worsen burns. Do not induce emesis, since this exposes the oesophagus to more of the agent and do not attempt to pass a nasogastric tube or perform gastric lavage as they can cause oesophageal perforation. Activated charcoal is ineffective.

Patients who are ill or have severe mouth burns at presentation require close observation and early intervention. Assess the airway carefully and seek anaesthetic/ITU advice very early. It is best to intubate these patients before marked pharyngeal swelling develops, this may be extremely difficult and requires anaesthetic expertise. Give supplemental oxygen, obtain IV access, begin fluid resuscitation if in shock, give analgesia and anti-emetics if needed. Obtain fbc, U&E, ABG, ECG, and a chest X-ray. These patients are usually managed in an ITU with surgical review.

Patients who are well with nothing more than pharyngeal irritation require admission for oesophagoscopy. Obtain IV access, baseline investigations, and ensure close observation.

Organophosphate insecticides

These agents vary widely in toxicity and pharmacokinetics. All kill insects by competitive inhibition of cholinesterase enzymes, and have the same effect in humans, resulting mainly in muscarinic parasympathetic effects, due to a post-synaptic excess of acetylcholine. The typical presentation is referred to as the 'SLUD' syndrome' with salivation, lacrimation, urination, and defaecation. Sweating with excessive bronchial secretions, bronchospasm, bradycardia, and miosis also occur. In severe cases, there may be cardiovascular collapse, muscle fasciculation, and paralysis.

Organophosphates are excreted through the skin so keep the patient in a well-ventilated area and do not touch contaminated clothing or secretions. Wash the patient thoroughly if skin contamination has occurred. Manage the patient with supportive measures, activated charcoal and also the following:

Atropine   Start with 0.5 mg IV and titrate until there are signs of full atropinization. Expect to require very high doses in serious poisoning (10–20 mg/h has been required).

Pralidoxime   This is a specific antidote effective against organophosphates. It is given in a dose of 1–2 g over 10 minutes and may be repeated. Pralidoxime should not be used for poisoning with carbamates, another type of acetyl-cholinesterase inhibitor insecticide. Discuss your patient with the poisons' information centre.

All patients with symptomatic organophosphate toxicity need to be managed on an ITU.

Ethylene glycol

Commonly available as antifreeze, this substance is metabolized by alcohol dehydrogenase to oxalic acid. This results in severe metabolic acidosis and renal failure due to oxalate crystalluria. Patients may appear intoxicated and hyperventilation is common due to metabolic acidosis. Poor peripheral perfusion and shock occurs in severely poisoned patients. Ethylene glycol levels can be measured and osmolality will be high.

Treatment can be started prior to a biochemical diagnosis. Treat with supportive measures. Adequate hydration should be maintained. After discussion with your ITU and the poisons' information centre, treat as for methanol poisoning with IV or oral ethanol to competitively block the alcohol dehydrogenase. The initial dose is 0.6 g/kg, then approximately 0.1 g/kg/h to maintain a serum ethanol of 22 mmol/l

Charcoal haemoperfusion and haemodialysis are effective at removing ethylene glycol. It is therefore essential that symptomatic patients are referred to the ITU early.

Methanol

No longer contained in 'methylated spirits', methanol is now seen as a by-product of illicit brewing, in antifreeze, or as an industrial solvent. It is metabolized by alcohol dehydrogenase to formaldehyde, then formic acid, which causes metabolic acidosis and blindness. This is often a latent period, prior to symptoms developing, of 12 hours or more, prolonged if ethanol has been consumed, as this is a partial treatment. As with ethylene glycol poisoning there is a metabolic acidosis and osmolality is increased. Serum methanol may be measured. Treat with supportive measures and with IV or oral ethanol, doses as above, to competitively block alcohol dehydrogenase. Discuss the patient with your ITU early as they may require haemodialysis, particularly if symptoms are present.

Paraquat

Paraquat is a herbicide, ingestion of small quantities of which can be fatal. It causes initial corrosive effects in the pharynx. Over the following 1–3 weeks there is progressive pulmonary damage, with an ARDS-type picture leading to pulmonary fibrosis and respiratory failure. This pulmonary damage is accompanied by a progressive multiorgan failure, particularly affecting hepatic and renal function. The toxic effects of paraquat are worsened by high concentration oxygen.

Because of paraquat's toxicity, if a patient presents early every effort must be made to remove it from the stomach. Perform gastric lavage with appropriate airway protection, and use repeated doses of Fuller's Earth plus mannitol (200 ml of 20 per cent solution), rather than charcoal, as an adsorptive agent. However, if only charcoal is available, do not delay and use that. Give the lowest inspired oxygen concentration that will maintain So_2 above 90 per cent. Discuss the patient with the ITU early.

Further reading

1. (1987) Poisoning with drugs and chemical substances. In *Oxford textbook of medicine*, (2nd edn), pp. 6.1–65. Oxford University Press.
2. Eilers, M. A. and Garrison, T. E. (1992). Toxicologic problems. In *Emergency medicine: concepts and clinical practice*, (ed. P. Rosen), pp. 2470–89. Mosby-Year Books, St. Louis, MO.

Environmental emergencies

Hypothermia

Hypothermia is defined as a core temperature less than 35 °C.

Predisposing factors

- Environmental exposure (e.g. unheated housing, immersion in cold water, or mountains in winter).
- Inability to move (e.g. fractured neck of femur, head injury, cardiovascular accident, intoxication, overdose).
- Inability to generate heat (e.g. hypothyroidism, hypopituitarism, malnutrition).
  In the elderly, hypothermia is often multifactorial.

Initial assessment

- **Assess ABC** The airway must be open. If the patient needs intubation for airway protection it should be done by an experienced person, usually an anaesthetist. Has the patient had a cardiac arrest? (see below for management).
- **If you suspect hypothermia:**
  - give oxygen and establish IV access,
  - measure the core temperature on a low-reading rectal thermometer. Check BM stix.

Further assessment

- **Full history** including:
  - did the patient fall or collapse? Any previous collapses? Any injuries?

- symptoms of myxoedema,
- previous medical, drug, and alcohol history. Empty containers suggesting overdose?
• **Full examination**:
- examine the patient a bit at a time covering the rest of the body—avoid making them colder,
- look for signs of injury (e.g. head injury or fractured neck of femur, pressure sores, or signs suggesting rhabdomyolysis).

Investigation

- Chest X-ray: for signs of infection or aspiration. Other X-ray if there is any suspicion of injury.
- ECG: for arrhythmias or signs of a recent myocardial infarction. May show J waves below 32 °C.
- Fbc, U&E, and glucose.
- Thyroid function tests can be done later by the in-patient team.

Treatment of mild/moderate hypothermia

This is a temperature above 30–32 °C.

Oxygen

IV access and cardiac monitoring because of the risk of arrhythmias. If IV fluid is needed for dehydration or maintenance requirements, it should be warmed.

Passive external rewarming Take off all the patient's clothes and wrap the patient in polythene, including the limbs and scalp. Put blankets over the polythene. Cover the scalp with blankets or a woolly hat. Aim for a temperature rise of 0.5–1 degrees per hour.

Arrhythmias Atrial fibrillation is common and usually resolves spontaneously with rewarming. As the patient gets colder ventricular fibrillation (VF) or asystole will supervene (see the cardiac arrest section below).

Avoid unnecessary movement as this may precipitate VF.

Active external rewarming with hot baths or electric blankets should be avoided in all but the young following brief immersion, as it may precipitate VF.

Pancreatitis is treated in the standard manner.

The role of antibiotics is controversial but if there you seriously suspect infection treat with broad-spectrum antibiotics after cultures have been taken.

Hypothermic patients require admission to a high-dependency unit.

Treatment of severe hypothermia

This is a temperature below 28–30 °C.

This carries a significant mortality Heat generation is impaired and active core rewarming is likely to be required.

Assess ABC ensuring the airway is clear, that the patient is breathing and has a cardiac output.

Give high concentration oxygen, obtain IV access, and start cardiac monitoring

Refer to the ITU for active core rewarming. These patients should be discussed with the ITU at an early stage, before instituting the invasive procedures listed below:

- Warmed humidified oxygen. The ITU physical will have access to the appropriate anaesthetic circuit.
- Warmed gastric, pleural, or peritoneal lavage.
- Cardiopulmonary bypass. This is the most effective but takes some time to set up and is not available in all centres.

Cardiac arrest

Start cardiopulmonary resuscitation (CPR). Resuscitation is likely to be prolonged and should be managed by a **senior** clinician. The patient should be intubated and drugs given through a central line. Passive external rewarming will not be effective in the absence of an effective circulation and active core rewarming, possibly with thoracotomy, and internal cardiac massage will be required (see above). In centres with access to cardiopulmonary bypass this is the treatment of choice. Resuscitation **must** be continued until the patient is warmed to 35 °C.

They're not dead 'til they're warm and dead.

Cautionary point

Every time you certify death think: 'Is this patient dead or only hypothermic?'

Heat illness

There is a spectrum of disease that arises from excess body heat. Heat is generated internally, mainly by muscle activity, and is lost into the environment, by evaporation of sweat and by conduction. Thus, the risk factors for heat illness are heavy exercise, inadequate fluid intake, and hot conditions. The worst cases tend to occur in semi-professional athletes in hot weather, but milder conditions are seen in the immobile elderly and the active young in hot conditions without water. Classical presentations occur in army recruits in training, marathon runs, and sometimes in epileptics after prolonged seizures. Recently, it has become a problem at 'raves' where the hot atmosphere combined with dancing, expensive drinks, and the use of drugs, such as Ecstasy, amphetamines, and cocaine, have caused many deaths.

Heat syncope

This is caused by the combination of dehydration and peripheral vasodilation producing hypotension and a simple faint. Thermoregulation remains intact. The condition responds to cooling by removal of clothing, removal from heat, rest, and oral fluids. By the time of presentation to hospital, most of these patients have improved and required support and fluids only.

Heat exhaustion

In this condition, thermoregulation remains intact but is overwhelmed by the environment. The patient has a raised core temperature (38–41 °C), significant dehydration and electrolyte depletion, but preserved cerebral function. These patients are still unwell on presentation to hospital, and require cooling and IV fluids, and usually admission. Symptoms include thirst, weakness, muscle cramps, nausea, and vomiting. Check for rhabdomyolysis and cardiac damage (U&E, CK, ECG, urinary myoglobin). Untreated, this may progress to heat stroke.

Heat stroke

This is a life-threatening medical emergency, produced when

hyperthermia has produced sufficient cerebral damage to cause breakdown of thermoregulation, and thus ongoing brain damage. Patients present with encephalopathy, severe hyperpyrexia (core temperature 41–43 °C), and usually hot, dry skin. The CNS symptoms range from mild confusion to convulsions and profound coma.

Immediate assessment and management

Airway May be at risk if comatose. Check and clear by whatever means necessary.

Breathing Initial hyperventilation is complicated by development of metabolic acidosis. Give high-flow oxygen and support breathing as necessary. If intubation is required, get senior help and *do not use suxamethonium* for paralysis (this worsens hyperkalaemia and hyperthermia).

Circulation The patient is in shock and has impaired coagulation due to disseminated intravascular coagulation. Obtain IV access and begin immediate resuscitation with cool (room temperature) normal saline. Take blood for fbc, coagulation profile, U&E, LFT, CK, myoglobin, and immediate blood glucose.

Neurology Control seizures and marked agitation with IV diazepam if needed, whilst continuing with cooling, which is the most effective treatment.

Cooling Immediately remove all the patient's clothing and begin cooling. The most effective method is to use a fan and to keep the patient moist by sponging or covering them in fine gauze and wetting it regularly. An easier but less effective method while this is being set up is to use ice packs in the groins, axillae, and neck. Monitor core temperature and cool as quickly as possible to 39°C. There will be a significant 'afterdrop' once cooling is stopped. Some units use cool peritoneal lavage or other invasive techniques of cooling: seek senior advice in your hospital.

Formal assessment Undertake history and examination, seeking underlying illness, precipitant, and cardiovascular, cerebral, or renal damage. Insert an indwelling catheter for ongoing fluid management, and obtain chest X-ray ECG, urinary myoglobin.

Disposition Heat stroke is a severe disease and normally requires treatment in an ITU setting.

Malignant hyperthermia

This rare condition arises as a complication of anaesthetic agents including inhalation anaesthetics, suxamethonium, and, rarely, amide local anaesthetics. There is believed to be a genetic predisposition through a cellular membrane defect. The signs are muscle rigidity, tachycardia, increased metabolic rate, with subsequent hyperpyrexia, pulmonary oedema, co-agulopathy, and rhabdomyolysis.

Call for senior anaesthetic help immediately. Treat malignant hyperthermia rapidly by ceasing the causative agent, hyperventilation on oxygen, and rapid cooling. After discussion with senior anaesthetic staff give dantrolene IV at 1 mg/kg/minute to effect or 10 mg/kg maximum. All patients must be admitted to an ITU.

Neuroleptic malignant syndrome

This is an atypical response to neuroleptic drugs (phenoth-iazines, butyrophenones) caused by dopaminergic blockade impairing hypothalamic thermoregulation. Patients present with gradual onset of hyperthermia, rigidity, akinesia, impaired consciousness, and tachycardia, and progress to the major complications of hyperthermia described above.

Carbon monoxide poisoning

Carbon monoxide (CO), produced from incomplete combustion, combines with haemoglobin, with 200 times the affinity of oxygen, to form carboxyhaemoglobin (COHb). It reduces oxygen delivery and at a cellular level paralyses the cytochrome oxidase system.

Presentation

Carbon monoxide poisoning occurs in several settings:

- Fires: often coexists with skin and airway burns and other toxic gas poisoning.
- Car exhaust fumes: a common method of suicide.
- Poorly ventilated heaters: May affect an entire household.

The symptoms it causes vary from non-specific malaise, nausea and headache to confusion, coma, arrhythmias and death. It may cause long-term neurological impairment. If CO poisoning is suspected, a COHb level should be taken and treatment instituted immediately.

Initial treatment

Airway In all patients the airway must be open and protected but there may be significant airway compromise due to burns or a reduced conscious level. If you suspect airway burns, even if the airway is open at present, contact an anaesthetist immediately.

Breathing Check for breathing, if the patient is apnoeic start ventilation: 100 per cent (O_2) oxygen must be given to all patients. It increases the amount of dissolved O_2 in the blood, increasing oxygen delivery, and reduces the half-life of COHb, limiting its toxicity: 100 per cent O_2 can be delivered through an anaesthetic facemask connected to an anaesthetic circuit, or, until this is available, 85 per cent O_2 can be given through a reservoir mask connected to 15 l/min O_2. Remember that the pulse oximeter records total saturated Hb (i.e. oxy-met-, and carboxyhaemoglobin). Your patient may have a pulse oximeter reading 100 per cent but be profoundly hypoxic.

Circulation Check for a pulse, if the patient has suffered a cardiac arrest cardiopulmonary resuscitation should be started. All patients require iv access and cardiac monitoring because of the risk of arrhythmias. Cardiogenic shock with severe metabolic acidosis is common. This requires urgent referral to the ITU for invasive monitoring and inotropic support.

Investigation

COHb The COHb level should be measured as soon as possible. It is expressed as a percentage of total Hb. Because of CO's cellular toxicity the COHb level does not correlate well with toxicity. Smokers may have a level up to 10 per cent, levels over this suggest significant exposure, and levels over 25 per cent can be associated with severe toxicity.

Arterial blood gases As pulse oximetry is useless in the presence of COHb, ABGs must be done to assess oxygenation and the severity of acidosis. Even if the PCO$_2$ is satisfactory,

100 per cent O_2 must be given until all the CO has been cleared.

Chest X-ray Smoke inhalation can result in immediate or delayed pulmonary oedema. A chest X-ray should be taken on admission.

Blood glucose Many of the symptoms of CO poisoning resemble hypoglycaemia.

ECG In all patients with significant CO toxicity a 12-lead ECG should be performed, because of the risk of myocardial ischaemia and infarction.

In ill patients, **baseline fbc and U&E** should be performed

Hyperbaric oxygen The use of hyperbaric oxygen for CO poisoning is controversial and there is a lot of variation in its use. There are anecdotal reports of good results but there are no randomized controlled trials that confirm its benefit. A compromise used by some is to reserve its use for more severe poisoning (i.e. COHb >25 per cent) or persisting neurological abnormality. Fetal Hb has a higher affinity for CO than adult Hb so there should be a lower threshold for hyperbaric O_2 use in pregnancy.

If you have a patient you feel might benefit from hyperbaric O_2 discuss the case with your consultant and the medical staff at your local chamber. Usually, patients need to be transferred some distance for hyperbaric therapy so ensure that the transfer is safe with appropriate equipment and an experienced medical escort.

Decompression sickness (the bends)

Divers with decompression sickness (DCS) can present at any time of year to any hospital, on the coast or inland. DCS can mimic many other conditions but if not recognized and treated correctly the results can be devastating.

Pathophysiology

With the increased pressure underwater, more nitrogen dissolves in the blood and the tissues than at the surface. If the dive is very deep or prolonged, the ascent is rapid or a diver flies or climbs a mountain within 24 hours of a dive, bubbles of gaseous nitrogen will form in the tissues and the circulation. DCS occurs when the bubbles cause symptoms.

Presentation

DCS usually presents within a few hours of a dive but can present up to 36 hours later. Ask the diver about his/her recent dive profiles. It is possible to develop DCS when diving within dive table or computer 'safe' limits. Enquire about his/her dive buddy as they may also have DCS. The presenting symptoms and signs depend on where the bubbles have formed:

Lungs Chest pain, cough, or breathlessness.
Spinal cord Back pain, limb parasthesia, weakness, paralysis, or urinary retention.
Brain headache, visual symptoms, any focal deficit, or abnormal conscious level.
Muscles/joints—the most common presentation of DCS Aching, numbness.
Skin rashes, mottling, and pain.

Even if the symptoms are restricted to the skin or a small joint there may be other bubbles in the brain or spinal cord so all patients need full treatment including recompression.

Treatment

Assess and treat ABC.
Oxygen All patients with DCS must be given as high a concentration of oxygen as is possible. Increasing the FiO_2 decreases the amount of nitrogen in the alveoli, increasing nitrogen elimination, and improves oxygen delivery to poorly perfused tissues, minimizing the amount of ischaemic injury.
IV fluids Begin volume expansion. Dehydration commonly follows diving and will worsen DCS.
Analgesia DCS can cause severe pain which should be controlled by IV opiates. **Do not give entonox** as nitrous oxide will diffuse into the gas bubbles, making them larger.
Recompression Recompression is the definitive treatment for DCS and **must** be performed in all cases. The increase in pressure redissolves the bubbles.

If you even suspect DCS, phone and speak to an expert. The

telephone number of your local recompression chamber and the national telephone advice line should be displayed in your A&E department. Transfer to the recompression chamber should be by road if possible. If aerial transfer is unavoidable due to the distance involved or the terrain, the aircraft should fly as low as is safe, preferably below 153 metres (500 feet).

Envenomation

Envenomation by bees, wasps, and ants is common, although only those with significant symptoms, usually allergic in origin, present to hospital. Envenomation by insects or animals with major toxicity is comparatively rare, and, since the development of antivenom and first aid techniques, deaths are extremely uncommon.

First aid

Envenomation or potential envenomation with any rapid-acting systemic venom except box jellyfish toxin (see below) should be treated with a pressure bandage and immobilization. In Australia, this includes all snakes (including sea snakes), the funnel-web spider, and the blue-ringed octopus. It does not include the redback spider, stonefish, nor other fish, which cause local pain of slow onset, nor the North American snakes. Apply a wide bandage to the whole limb, starting at the bite site, as firm as for a sprained ankle, and then bandage to a splint.

Box jellyfish stings must be treated with immediate application of vinegar to prevent further envenomation by adherent tentacles before pressure immobilization is applied. All other fish envenomations should be treated by dipping the affected part in a bucket of hot water (as hot as the patient can stand) for 30 minutes since the toxic elements are heat-labile proteins.

Immediate management of serious envenomation

Patients who appear systematically unwell after any envenomation must be treated rapidly and aggressively. If first aid has not been given, then give it whilst undertaking assessment. The effects vary between toxins, but all are treated with supportive measures and antivenom if available (none is yet available for the blue-ringed octopus) Snake venoms contain a mixture of neurotoxins (causing paralysis), myotoxins (causing rhabdomyolysis), and procoagulants (causing disseminated intravascular coagulation); and envenomation produces secondary shock and renal failure. Funnel-web spider venom causes dramatic autonomic stimulation due to release of endogenous transmitters with tachycardia, hypertension, pulmonary oedema, muscle spasms, and acidosis. Ticks, which may be well hidden under the skin, produce slow motor paralysis, which can be mistaken for neuropathy. Blue-ringed octopus envenomation causes skeletal muscle paralysis. Box jellyfish stings cause extreme pain, paralysis, and hypotension which may rapidly progress to death. As in all medical emergencies, supportive treatment comes first:

Airway Check for a patent airway and open the airway by whatever means are necessary.

Breathing If not breathing, begin ventilation with bag-valve-mask, and prepare for endotracheal intubation. If breathing, give high-flow oxygen by facemask, and assess for hypoxia and muscle power (paralysis may be worsening).

Circulation If there is no cardiac output, then commence cardiopulmonary resuscitation and treat as for cardiac arrest (see Chapter 2). Obtain IV access and begin fluid resuscitation if in shock.

First aid Check again whether first aid was properly administered and if not, give it now.

History and examination Undertake formal assessment to determine the organism involved, number, and location of bites or stings, symptoms, and signs of envenomation. Find out if the patient has had bites or antivenom before.

Investigation Obtain fbc, U&E, coagulation studies, plus serum and urinary myoglobin if unwell, and for snake bites, a sample for identification by a venom detection kit. This is preferably a washing from the bite site (it is acceptable to cut a window in the bandage) or urine sample.

Antivenom therapy Antivenom is indicated in patients with significant symptoms (i.e. more than minimal headache or pain) or laboratory abnormalities (detection of venom in the urine alone is not sufficient). The dose to be given is venom-related, not patient-related, so do not adjust for patient size. Full instructions on antivenom use are beyond the scope of this book. Read the product information carefully, and be cautious in patients who have had previous doses of antivenom, since allergic reactions are more common. In most cases, it is appropriate to pretreat with an antihistamine, such as chlorpheniramine 10 mg IV, and possibly hydrocortisone 100 mg IV. Resuscitation equipment and adrenaline should always be close at hand.

Removal of first aid   There are anecdotal reports of sudden onset of severe symptoms after removal of pressure immobilization following snakebite. Hence, this first aid should not be removed until after arrival in hospital. In patients with systemic symptoms, it should not be removed until IV access, laboratory tests, venom identification, and appropriate antivenom are available. If there is evidence of significant envenomation, antivenom should be given before removal.

Disposition Patients who have had serious envenomation must be admitted to hospital. Those with ongoing problems should be treated in the ITU.

Severe local envenomation

Although not life-threatening, except in small children, the redback spider and stonefish cause severe local pain with some systemic symptoms. There is risk of ongoing tissue damage without treatment. Again, antivenom is indicated if symptoms are significant. It may be effective even days after the sting. Patients with a short history who respond to antivenom may be discharged after a few hours observation, but severe cases require admission to hospital.

The well patient after envenomation

The majority of patients who present after bites as well. Most will be able to be discharged after observation. For snakebites, obtain IV access, and check fbc and coagulation profile. Remove the pressure immobilization bandage, and then observe for 4 hours or to a total of 6 hours after the bite. For snakebites, check coagulation profile prior to discharge, but for all other bites, discharge if asymptomatic after observation.

Further reading

1. Steedman, D. J. (1994). *Environmental medical emergencies*. Oxford University Press.
2. Edmonds, C., Lowry, C. and Pennefather, J. (1992). *Diving and subaquatic medicine*. Butterworth-Heinemann, Oxford.
3. Lippman, J. (1990). *Deeper into diving*. JL Publications, Australia.
4. Hoffman, R. S. (1992). Toxic inhalations. In *Emergency medicine: concepts and clinical practice*, (ed. P. Rosen), pp. 2673–84. Mosby-Year Books, St. Louis, MO.
5. Danzl, D. F. (1992). Accidental hypothermia. In *Emergency medicine: concepts and clinical practice*, (ed. P. Rosen), pp. 913–44. Mosby-Year Books, St. Louis, MO.
6. Yarbrough, B. (1992). Heat illness. In *Emergency medicine: concepts and clinical practice*, (ed. P. Rosen), pp. 944–64. Mosby-Year Book, St. Louis, MO.
7. Otten, E. J. (1992). Venomous animal injuries. In *Emergency medicine: concepts and clinical practice*, (ed. P. Rosen), pp. 875–93. Mosby-Year Books, St. Louis, MO.

CHAPTER 11

Haematological emergencies

Profound anaemia

The effect of anaemia depends on the rapidity of its onset and the ability of the patient's cardiovascular system to compensate for it. Any patient who has symptoms at rest due to anaemia should be given high concentration oxygen, while awaiting transfusion. The need for transfusion in an anaemic patient depends not on the degree of anaemia, but on the patient's clinical state and the risk of a continuing fall in the haemoglobin concentration.

Box 11.1 Causes of anaemia

- Reduced red blood cell production:
  Iron, folate, or vitamin B_{12} deficiency, marrow infiltration, chronic disease, thalassaemia
- Increase red blood cell loss:
  Bleeding, haemolysis (including sickle-cell disease)

Blood loss

The most common cause of profound anaemia presenting to the A&E department is blood loss, which may be chronic, acute, or a combination of both. It is important to realize that the degree of anaemia does not reflect the amount of acute blood loss. A patient with active bleeding will not have any

fall in haemoglobin until there has been time for haemodilu-
tion. All patients with significant acute blood loss require oxygen and IV access, but volume resuscitation should be based
on evidence of hypovolaemia, such poor perfusion, not the
haemoglobin concentration.

Anaemic patients thought to be at risk of ongoing bleeding should be transfused, even if they are not compromised
by the anaemia, as hypovolaemia in a patient who is only
just compensating for reduced oxygen transport, may be devastating.

Investigation

It is essential that a precise diagnosis is made in patients with
anaemia so that it can be correctly treated and future episodes
prevented. Even in the few patients who require urgent transfusion because of haemodynamic compromise, blood should
be taken from the cannula for the following tests, prior to
transfusion being started, as transfusion can affect the results:
group and cross-match, fbc and film, reticulocyte count, and
U&E. Take blood into an extra EDTA and clotted tube and
label them with the patient's details so that other investigations can be arranged by the in-patient team, these include:
iron, total iron-binding capacity (TIBC), ferritin, vitamin B_{12},
folate, and Coomb's test.

Other tests that may need to be performed in patients where
the cause for the anaemia is not obvious include: thyroid function tests, liver function tests, haemoglobin electrophoresis,
and haptoglobin. Bone marrow examination may also be
required as a planned procedure, on the ward.

Indications for admission

Anyone with acute blood loss, or who may have acute
haemolysis needs urgent admission.

In patients with chronic anaemia, presenting to the A&E
department, indications for admission include: symptoms,
such as dyspnoea; chest pain or neurological symptoms; or a
haemoglobin concentration below 9 g/100 ml. Patients with a
lesser degree of anaemia require investigation and treatment,
but it may be appropriate to discuss them with the medical
team so that this can be arranged as an out-patient.

The immunocompromised patient

It is not uncommon for immunocompromised patients to present to the A&E department. Treatment with cytotoxic agents, immunosuppression following transplantation, and AIDS are obvious causes of an immune deficit but remember that diabetic patients and those following splenectomy or on steroid therapy are also immunocompromised.

Bacterial infections must be high in the differential diagnosis of any ill immunocompromised patient. As they are unable to mount the normal physiological responses that usually make us suspect infection, such as pyrexia and an elevated white blood cell count, diagnosis may be difficult and empirical treatment may have to be given. As an example, without a normal white blood cell response, pneumonia will not be associated with purulent sputum or an elevated white blood cell count, and the physical signs of consolidation may not occur.

Fever in the neutropenic patient

There are many causes for fever in patients who are neutropenic, including drug reactions and tumour necrosis, but the most common cause is bacterial infection. These patients may be critically ill and it is essential to start treatment immediately, before confirming a bacteriological diagnosis.

Resuscitation All critically ill patients should be triaged to the resuscitation room. Ensure the airway is clear and give high-flow oxygen, connecting the patient to a pulse oximeter. Obtain IV access, and if the patient has signs of shock start fluid resuscitation (see Chapter 12).

Investigation Take blood for fbc, U&E, glucose, and blood cultures. If there are any signs of respiratory compromise of So_2 is low take arterial blood for acid–base analysis. Arrange a chest X-ray.

Management Empirical, broad-spectrum antibiotics should be started in the A&E department, after obtaining cultures. A common antibiotic regimen in this setting is piperacillin and gentamicin, but discuss the patient with the oncology team before instituting therapy.

Patients on chronic steroid therapy may have adrenal suppression and therefore be unable to mount a normal steroid response to serious illness. Give an additional dose of steroid, (e.g. hydrocortisone 100–200 mg IV), to prevent an Addisonian crisis. Many modern chemotherapeutic regimes result in good cure rates so never assume that because a patient is receiving chemotherapy they would not be suitable for ITU care. The patient will be well known to the oncology unit so continue maximal resuscitation while you contact a senior member of the oncology team.

Respiratory problems

Although patients with immunocompromise are at risk of atypical infections, particularly *Pneumocystis carinii* and cytomegalovirus (CMV); they are also more prone to simple bacterial infections than the general population. In severely immunocompromised patients, the classical symptoms, signs, and chest X-ray appearances of pneumonia may not occur and the patient may present with a 'septic' picture with little to suggest the site of the infection.

Patients presenting with sudden respiratory failure require high-flow oxygen and often ventilatory support. As discussed above, it is not appropriate for a doctor who does not know the patient to decide to withhold resuscitation or ITU facilities. Seek expert help, start full resuscitation, and contact a senior member of the patient's medical team to discuss further management.

Patients with pneumocystis pneumonia may have a dry cough and a feeling of breathlessness with little to find on examination or chest X-ray, despite severe hypoxia. Worsening breathlessness should never be ignored. All such patients require investigation.

Mycobacterial infection, due to both *Mycobacterium tuberculosis*, and atypical mycobacteria are more common in immunocompromised patients than in the general population. Radiological appearances may be atypical. If the diagnosis is suspected, sputum can be sent for a rapid stain for acid-fast bacilli, but a more invasive investigation may well be required. All such patients should be referred for admission by their own oncology team to the infectious diseases or respiratory unit.

Neurological problems

As well as the neurological conditions that affect the general population, patients who are immunocompromised are prone to other neurological diagnoses, such as *Listeria* or cryptococcal meningitis, cerebral toxoplasmosis, and cerebral lymphoma. Patients who do not have an intact immune system who present with neurological symptoms or signs require admission and investigation. HIV encephalopathy may present with a first fit or an alteration in mental state but opportunistic infection and lymphoma must always be excluded.

What about reverse barrier nursing?

Severely neutropenic patients who need to be in hospital should be admitted to a specialist oncology unit, where they can be in a single room with reverse barrier nursing, to prevent further infection. It is impossible to practise barrier nursing in the A&E department.

AIDS

The incidence of HIV infection and AIDS is continuing to increase world-wide and in many areas heterosexual intercourse is the most common source of new infection. Many patients are only found to be HIV-positive when they present, severely ill with their first opportunistic infection, so AIDS should be considered in the differential diagnosis of many conditions.

If a patient with AIDS requires admission, whenever possible arrange for them to be admitted to the unit responsible for patients with HIV—usually infectious diseases or genitourinary medicine. They have ongoing medical problems, for which they need both medical expertise and emotional support.

Pneumonia

Patients with AIDS may develop pneumonia due to pathogens that commonly affect immunologically normal patients, such as *Streptococcus pneumoniae* and *Haemophilus influenzae*, but they are also prone to infections with unusual organisms such as *Pneumocystis carinii*, CMV, and mycobacteria. In many parts of the world, the increase in HIV infection has been paralleled by an increase in the incidence of pulmonary tuberculosis. In the United States, the development of multiresistant strains of *Mycobacterium tuberculosis* is causing major concern.

It is important to take the symptom of unexplained breathlessness very seriously as patients with *Pneumocystis pneumoniaa* may have a normal chest X-ray and few abnormal physical signs, despite marked breathlessness and hypoxia. If you suspect pneumonia, clinically or radiologically, the patient will need to be admitted for further investigation, which may include examination of induced sputum or bronchoalveolar lavage. It is unusual to be able to make a microbiological diagnosis in the A&E department. Give supplemental oxygen to any patient who is breathless at rest or who has a low oxygen saturation on pulse oximetry.

Neurological problems

AIDS may lead to neurological symptoms and signs due to the direct effects of HIV infection, opportunistic infection (e.g. *Toxoplasma gondii, Cryptococcus neoformans, M. tuberculosis*, and *Listeria*) or neoplasia, either primary or secondary. Common acute presentations include seizures, focal neurological deficits, and altered mental state.

Supportive treatment with airway management and control of seizures may be required immediately. Confirmation of the cause of the neurological problem usually requires a CT scan of the brain. If there are no contraindications to lumbar puncture seen on the CT scan, examination of the CSF may enable a specific microbiological diagnosis to be made.

Any HIV-positive patient with new or changing neurological symptoms or signs will need admission to hospital for further investigation.

Sickle-cell disease

Patients with sickle-cell disease or trait may present to the A&E department with complications of their conditions, but may also present with coincidental illness or injury. It must be remembered that some normal A&E department procedures, such as general or regional anaesthesia, will need to be modified in sickle-cell disease. Any patient of Afro-Caribbean origin in whom anaesthesia is contemplated, should have a sickle-cell test performed. This should not delay emergency anaesthesia but if the result is not available anaesthesia should be performed as if the result were positive.

Pathophysiology

In sickle-cell disease a single amino acid change in the haemoglobin molecule transforms HbA to HbS. This change results in the haemoglobin crystallizing in the deoxygenated state, causing a deformed or 'sickled' cell. These cells are rigid and tend to sludge in the circulation, causing vessel obstruction. There is chronic haemolysis and vasoocclusive episodes and thrombosis result in organ injury.

In sickle-cell trait half the haemoglobin is HbS and half normal HbA. It is usually asymptomatic, although sickling may occur when the oxygen tension is very low, such as at altitude, or if hypoxia occurs during anaesthesia.

Sickle-cell disease occurs mainly in the Afro-Caribbean population but a small number of cases occur in people originating in the Mediterranean, the Middle East and India.

Painful crisis

These may occur at any site. They may both resemble and coexist with acute infection. They are thought to occur because of tissue ischaemia due to sludging in the microcirculation. Continued sickling occurs within the hypoxic tissues. Painful crises may be precipitated by infection, trauma, or exposure to cold, but frequently occur without any obvious cause.

There is a severe, dull, aching pain, most commonly in the abdomen, back, chest, or limbs. The pain may mimic many other conditions, such as cholecystitis or renal colic, causing diagnostic problems. If the differential diagnosis includes an acute abdomen the patient should be reviewed jointly by the haematology and surgical teams.

Treatment of a painful crisis consists of relieving symptoms and attempting to break the cycle of sludging and sickling. Contact the patient's haematology team for advice as soon as possible.

Oxygen   Supplemental oxygen should be given by facemask.

Analgesia   Most patients are in severe pain and analgesia is required, usually an opiate such as morphine. A small proportion of patients may be drug-seeking but if in doubt, it is better for the occasional patient to get a free dose of opiate than to leave a patient in agony, without analgesia.

Hydration   In sickle-cell disease there is loss of renal concentrating ability. Dehydration is common in painful crises and increases the tendency to sludging. Rehydration with 4 per cent dextrose 0.18 per cent saline should be started. Adults with sickle-cell disease are at risk of cardiac failure and hydration must be closely monitored.

Antibiotics   Infection is a common precipitant of painful crises and many haematology units routinely give antibiotic cover, with an antipneumococcal antibiotic such as penicillin. Discuss this with your patient's unit.

Infection

Patients with sickle-cell disease have an increased susceptibility to infection, due to reduced splenic function following multiple infarcts. Overwhelming infection with pneumococcus or staphylococcus may occur and there is an increased risk of *Salmonella osteomyelitis*.

Haemophilia

Haemophilia A (or classical haemophilia) is a condition in which factor VIII is abnormal, with reduced function. The severity of the clinical manifestations depends on the degree of activity of the factor VIII. The common sites for bleeding are muscles and joints but urinary tract and intracranial bleeding may also occur. Patients with haemophilia are usually very well informed about their condition. It is essential to pay attention to what they have to say, they are often aware of a haemarthrosis developing before there are any objective physical signs. Whenever a patient with haemophilia presents with actual or potential bleeding they should be discussed with the haemophilia centre.

Because of the past use of unscreened cryoprecipitate,

blood-bourne infection, including HIV and hepatitis C, is common in patients with haemophilia. This need only affect your approach to the patient by making you more alert to the possibility of opportunistic infection. Blood and body fluids from every patient you treat must be assumed to be infective.

Active bleeding

If a patient presents with active bleeding a rapid assessment should be made of their cardiovascular status to see if resuscitation is required. In most cases, although the bleeding may cause severe local problems it is not sufficient to result in hypovolaemia.

Minor abrasions and superficial lacerations can be managed by standard wound care, local pressure, and reassessment. More major injuries or spontaneous bleeding into muscles, joints, or urinary tract will need treatment with factor VIII. Discuss this with the patient's haematology team. Analgesia should not be forgotten.

Trauma

There are some situations where there is a risk of delayed bleeding. Patients with deep lacerations, soft-tissue bruising in areas where a large haematoma would be disastrous, such as the eye, neck, or spine, and those who have suffered significant deceleration forces but appear uninjured should be considered for factor VIII and admission.

Even apparently minor head injury can result in intracranial bleeding in a patient with haemophilia. All patients should receive factor VIII and be admitted, and all but the most minor should have a CT scan performed.

Von Willebrand's disease

In this condition, there is a reduced level of factor VIII and platelet dysfunction. The clinical manifestations may occasionally be similar to haemophilia A but they are usually less severe, with predominantly mucosal and cutaneous bleeding. Treatment of active bleeding, after discussion with the haemophilia centre, is with factor VIII.

Christmas disease (or haemophilia B)

This is due to a deficiency in factor IX activity. It causes a clinical syndrome identical to haemophilia A, but is less common. Treatment is with IV administration of clotting factors, after discussion with the haemophilia centre.

Further reading

1. Oh, T. E. (1990). *Intensive care manual.* Butterworths, Sydney.
2. Hamilton, G. C. (1992). Anemia and white blood cell disorders, and disorders of hemostasis and polycythemia. In *Emergency medicine: concepts and clinical practice*, (ed. P. Rosen), pp. 1673–707 Mosby-Year Books, St. Louis, MO.
3. Kelen, G. and Marco, C. A. (1992). AIDS and HIV-1. In *Emergency medicine: concepts and clinical practice*, (ed. P. Rosen), pp. 2311–27 Mosby-Year Books, St. Louis, MO.

Shock and fever

Hypotension

Shock is defined as an inadequate oxygen supply to the tissues. Hypotension may only develop in the advanced stages of the shock process. Other physical signs of shock include reduced urine output, cool peripheries (except in some cases of early septic and anaphylactic shock), and later, impaired cerebral function, ranging from mild confusion to coma.

When considering the blood pressure in a patient who may be shocked, absolute levels of systolic or diastolic blood pressure are less important than comparison with previous measurements and the response, or lack of response, to interventions. The causes of unexplained hypotension are listed below. The most common causes that are not suspected on initial presentation are probably gastrointestinal bleeding, 'silent' myocardial infarction, and occult sepsis.

Causes of unexplained hypotension

Reduced preload
Bleeding Gastrointestinal, occult trauma, ruptured abdominal aortic aneurysm.
Plasma loss Pancreatitis, gastroenteritis, diabetic ketoacidosis.
Obstruction Tension pneumothorax or pulmonary embolus.
Cardiac causes
Ischaemia and infarction causing impaired contraction.
Arrhythmias Tachy- or bradyarrhythmias.

168

Structural problems   Acute valve lesion or septal rupture due to infarction or infection.
Toxins   Iatrogenic or following self-poisoning.
Infection   Myocarditis and cardiomyopathy.
Cardiac tamponade.
Decreased afterload.
Anaphylaxis, sepsis, or vasodilators such as antihypertensive agents.

Patients may present to the A&E department, obviously unwell, with signs of shock, including hypotension, without any obvious cause for their illness. The patient may be unable to give a clear history. As in other areas of emergency medicine it is essential to start resuscitation while considering causes that require immediate treatment.

Immediate assessment and management

Obtain senior help early in the management of a patient in shock.
Airway—check that the airway is clear, and open it if it is obstructed.
Breathing if the patient is breathing give high-flow oxygen, preferably through a reservoir mask, if the patient is apnoeic ventilate using a bag-valve-mask and prepare for intubation.
Circulation—if the patient is unresponsive and there are no palpable pulses start cardiopulmonary resuscitation and treat as for cardiac arrest. If there are palpable pulses insert a wide-bore IV cannula and start IV normal saline or colloid such as gelofusin. Monitor ECG, BP, and So_2
Seek and start treatment of an easily reversible cause:
Hypoxia   Every patient in shock needs high concentration oxygen
Hypovolaemia   Give an initial fluid bolus of 10 ml/kg colloid (e.g. gelofusin), or 20 ml/kg crystalloid (e.g. normal saline or Hartman's solution). If there is no improvement repeat and call for expert help. If there is evidence of blood loss arrange an urgent cross-match and call for a member of the surgical team.
Arrhythmia   Check the ECG monitor and arrange a 12-lead ECG
Blood glucose   Check a bedside test and a formal blood glucose.

Hypoglycaemia may be contributing to the patient's symptoms. If hyperglycaemia is present consider diabetic ketoacidosis.

Fluid load—in all but patients with evidence of pulmonary oedema, an increase in circulating blood volume should improve cardiac output by increasing preload. Start a rapid bolus of 10 ml/kg of colloid (e.g. gelofusin) and assess its effect, while taking a history and examining the patient.

History—obtain a rapid but thorough history, seeking any localizing symptoms. Check the past history for clues such as previous gastrointestinal bleeding, a bleeding disorder, diabetes, or splenectomy. Ensure you know the patient's current medication: non-steroidal anti-inflammatory drugs frequently cause gastrointestinal bleeding and antihypertensive agents may cause profound hypotension.

Examination—after assessing ABC and starting resuscitation with oxygen and fluids, quickly perform a general examination. Visualize the whole body, looking for signs of bleeding. Always perform a rectal examination, to check for melaena.

Investigation

Be guided by your examination findings, but generally obtain, fbc, U&E, and glucose. Obtain a chest X-ray and 12-lead ECG. Cross-match blood if there is any suggestion of blood loss and take blood cultures if you suspect sepsis. A clotting screen may be required.

Definitive treatment

If you have a diagnosis start appropriate treatment (see relevant chapters). If you do not know the diagnosis, particularly if the patient has not responded to your initial fluid load get senior help rapidly. If there is evidence of ongoing blood loss call the surgical team. Any patient who has been shocked requires admission to a high-dependency unit or ITU with facilities for invasive haemodynamic monitoring. Clinical examination is poor at determining the haemodynamic status in shock. It is only by knowing the left atrial pressure, the systemic vascular resistance and cardiac output that fluid resuscitation and inotropic support can be tailored to the patient's needs. Early referral to the ITU is therefore essential.

Anaphylaxis

Anaphylaxis is the name given to the syndrome that results, when, in response to exposure to a substance to which an individual has been sensitized, there is mast cell degranulation and histamine release. This type 1 hypersensitivity reaction is accompanied by a massive release of other mediators, including leukotrienes, prostaglandins, and bradykinins. Common precipitants are drugs, such as antibiotics, foodstuffs, particularly nuts, and insect stings.

Shock occurs when massive histamine release results in circulatory collapse. Other manifestations include laryngeal swelling and obstruction, bronchospasm, an urticarial skin rash, and gastrointestinal symptoms including pain, vomiting, and diarrhoea. Any of these may occur in isolation.

Management

Call for senior help early.

1. Check for cardiac arrest. If the patient has arrested start cardiopulmonary resuscitation. Ventilate with 100 per cent oxygen and follow the Resuscitation Council guidelines (see Chapter 2). Start fluid resuscitation with 10–20 ml/kg colloid, stat. If there is no response, repeat the fluid bolus.

2. Oxygen. High concentration oxygen should be given to all patients with anaphylaxis.

3. Large bore peripheral IV access should be obtained and volume resuscitation started with a colloid such as gelofusin (unless this was the cause of the anaphylactic reaction), or a crystalloid (normal saline or Hartman's solution): 1–2 litres should be given initially and the patient should then be reassessed. Large volumes may be required, because of the peripheral vasodilation that occurs. Ensure the patient is connected to an ECG monitor.

4. Adrenaline. This is the drug of choice for anaphylaxis associated with airway compromise, bronchospasm, or cardiovascular collapse. It should be given IV, to ensure absorption, diluted, and carefully titrated against the response. Adrenaline 1 mg in 10 ml (1 in 10 000) should be given in 0.5 ml (50 microgram) aliquots, into a fast-flowing iv line. This improves cardiac contractility, peripheral vascular resistance, and reduces bronchospasm. Continue giving 50 microgram boluses until the patient improves. Adrenaline has a short half-life and an infusion may be required following the initial bolus. Seek advice on this from senior A&E or ITU staff.

5. The airway. Laryngeal oedema may compromise or totally obstruct the airway. IV adrenaline, as described above, should reverse this. If the airway totally obstructs, attempt intubation, if the patient is unconscious. Intubation will be extremely difficult, and if it is impossible (e.g. in a patient with total airway obstruction), a surgical airway will be required (see Chapter 16).

6. Other drugs. There is no evidence that steroids or antihistamines have any immediate effect in severe anaphylaxis, although they can be useful in preventing or ameliorating later deterioration. Suggested drugs and doses are hydrocortisone 100–200 mg, and chlorpheniramine 10–20 mg, slowly IV. Aminophylline or nebulized beta agonist can be given to patients whose bronchospasm persists despite adrenaline. All patients with anaphylactic shock must be admitted to an ITU or high-dependency unit. Even if the symptoms respond rapidly to treatment they may recur, and constant close observation is required.

Septic shock

Diagnosis

The diagnosis of septic shock can be difficult. The 'classical' presentation with a vasodilated, warm patient with a high fever and low blood pressure does occur but a large number of patients, particularly those presenting late, are peripherally vasoconstricted and have a normal or low core temperature. Tachypnoea is common and a purpuric skin rash may occur due to meningococcaemia or disseminated intravascular coagulation (DIC). Suspect the diagnosis in any ill patient with an altered level of consciousness or haemodynamic abnormality without an obvious cause. It is important to have a high index of suspicion.

Pathophysiology

The pathophysiology of septic shock is complex and not fully understood. The clinical signs are due to the patient's response to sepsis and are the same for Gram-positive and Gram-negative organisms, making it impossible to determine the causative organism from the physical signs.

Cytokines are a group of small protein inflammatory mediators that are produced by many cell types in response to the infecting organism's endo- and exotoxins. It is thought they may be beneficial when acting in low concentrations at a local level, but when there is massive cytokine production, the high concentrations in the circulation are thought to lead to the physical signs of septic shock, listed below.

In septic shock there is peripheral vasodilation which results in a dramatic fall in systemic vascular resistance, and hence blood pressure. Capillary leakage occurs, fluid and protein are lost from the circulation causing a relative hypovolaemia, and there is a reduction in venous tone leading to pooling of blood and a reduction in preload. In severe sepsis there is also myocardial depression, due to circulating myocardial depressant factors, hypoxia, acidosis, and myocardial oedema. This combination of reduced preload, poor myocardial contractility, relative hypovolaemia, and low systemic vascular resistance leads to cardiovascular collapse.

Resuscitation

Call for expert help early, as you start resuscitation.

Airway The airway must be clear and protected.

Breathing All patients require high concentration oxygen. Administration through a facemask may be adequate but if this does not restore normal oxygenation the patient will require ventilation. Seek anaesthetic help if you think ventilation may be required.

Circulation Obtain secure, wide-bore peripheral IV access. Start volume resuscitation with colloid (gelofusin, haemaccel, or human albumin solution), normal saline or Hartman's solution. Either 10 ml/kg of colloid or 20 ml/kg of crystalloid is an appropriate initial bolus in the shocked patient. Reassess after the first fluid bolus. By this time, expert help should be available. Repeated fluid boluses are likely to be required because of the loss of fluid from the circulation and the reduction in systemic vascular resistance, but ideally this should be done in an ITU with full haemodynamic monitoring.

If there is a purpuric rash or other features to suggest meningococcal septicaemia, after taking blood cultures, give benzyl penicillin 2.4 g IV immediately, without waiting for anyone more senior to arrive.

Investigations

Full blood count   Severe sepsis may occur with a normal or low total white cell count. There may be toxic granulation or left shift of the nutrophils and thrombocytopaenia can occur due to DIC.

Urea and electrolytes   Renal impairment is common.

Glucose In severely ill patients hypoglycaemia or hyperglycaemia may occur.

Clotting screen   DIC is common.

Arterial blood gases   These are essential to assess the severity of the acidosis and the presence of respiratory failure. Note the inspired oxygen concentration.

Blood cultures   Take several sets before commencing antibiotics.

Other cultures   Send urine, sputum and any pus for microscopy, culture, and sensitivity.

Chest X-ray   This may indicate the source of the sepsis or provide evidence of ARDS.

12-lead ECG.

Further management

Septic shock has a mortality rate of 50 per cent. All patients with septic shock will require admission to an ITU. Involve the ITU team early for advice.

Antibiotics   As the infecting organism will not be known, after several sets of blood cultures have been taken broad-spectrum antibiotics should be started. The particular antibiotic regimen will depend on the likely source of the infection, the patients' immune state, and the wishes of the in-patient or ITU team who will be responsible for the patient's ongoing management. A commonly used regime is a combination of a penicillin, gentamicin, and metronidazole. The only exception to this is suspected meningococcal septicaemia when benzyl penicillin 2.4 g IV should be given as soon as possible, after blood cultures, but, if necessary, before the ITU team arrive.

Monitoring While in the A&E department all patients require ECG and blood pressure monitoring and pulse oximetry. If there is no response to the initial resuscitation the patient will require monitoring of arterial, central venous and pulmonary artery pressures, and cardiac output measurement in the ITU.

Intropic support Administration of high concentration oxygen and volume loading constitute the initial management of septic shock. If patients do not respond to the increased preload from the initial volume resuscitation it may be that more volume is required or that inotropes are required to improve myocardial contractility and increase systemic vascular resistance. This decision, and the selection of an appropriate inotrope regime, can be made with the aid of invasive monitoring. In severely ill patients inotropic support may need to be started blind, prior to placement of a pulmonary artery catheter but this decision should only be made by senior, experienced staff.

Ventilation The indications for ventilation in septic shock are no different from those generally used in the ITU. Any patient who cannot protect their own airway requires airway protection and any patient who is unable to maintain normal Po_2 and Pco_2 values, who has a persisting metabolic acidosis, with elevated lactate levels, or continuing tachypnoea despite oxygen and volume resuscitation requires ventilation. If you suspect a patient may need ventilation contact the ITU.

Metabolic acidosis Lactic acidosis due to anaerobic metabolism is almost universal in septic shock. Its initial management consists of administration of high concentration oxygen and volume resuscitation, with inotropic support as necessary. Only if these fail to improve the acidosis, and after discussion with the ITU team, should sodium bicarbonate be considered.

Localized septic focus If a patient with septic shock has a localized, ongoing source of infection, such as gas gangrene or an intra-abdominal abscess, they should be referred to the surgical team.

Other drugs Steroids should not be used in septic shock.

There is no evidence that they benefit patients in septic shock, and it is thought they may have deleterious effects.

The future It is now recognized that the cardiovascular collapse in septic shock is due to the effects of inflammatory mediators. It is hoped that, in the future, specific antagonists may be available to prevent or reverse the haemodynamic changes of septic shock. Unfortunately, in early trials they have not resulted in any improvement in outcome.

Toxic shock

Toxic shock syndrome is thought to be due to toxin produced by *Staphylococcus aureus*. It most commonly occurs in menstruating women using vaginal tampons, but may occur in association with staphylococcal infection. Common symptoms are headache, myalgia, vomiting, and diarrhoea. Signs include a high fever (>38.9°C), a macular rash, which may desquamate, and hypotension. Multiorgan failure, including ARDS and DIC is common.

Management of suspected toxic shock syndrome

1. Ensure the airway is clear.
2. Give high concentration oxygen by reservoir mask.
3. Establish wide-bore IV access, and if there are signs of shock start volume resuscitation with 10 ml/kg of colloid. As in septic shock, large volumes of fluid may be required.
4. Any tampon should be removed and sent for culture.
5. Antibiotics are not thought to be of value in toxic shock syndrome but if septic shock is thought to be likely differential diagnosis broad-spectrum antibiotics should be started, once cultures have been sent (see p. 172)
6. All patients with suspected toxic shock syndrome require admission to an high-dependency unit or ITU.

'Foreign fevers'

With international travel, for business and recreation, being part of everyday life, any doctor may be faced with a patient who has recently returned from abroad. The differential diagnosis of fever following travel abroad is large and this section can only cover a few of the important causes. Few non-specialists are familiar with the epidemiology and patterns of drug resistance of these conditions. Discuss the patient with your local infectious diseases specialist or one of the schools of tropical medicine.

If a patient presents severely ill, even if you do not know the diagnosis, open the airway, ensure adequate ventilation with high concentration oxygen, and support the circulation with IV fluid if there are signs of hypovolaemia. This will keep the patient alive and prevent further deterioration until senior help arrives.

After assessing ABC, and starting resuscitation if required, take a full history, including a travel history. To formulate your differential diagnosis you will need to know all the areas the patient has visited, including airports, even if they did not stay overnight. Ask about immunization status and malaria prophylaxis. Sexually transmitted diseases are common after travel, enquire specifically about urinary and gynaecological symptoms.

In every patient who presents unwell after foreign travel consider common causes of their symptoms as well as the 'tropical' causes. Pneumonia, otitis media, and appendicitis should still be considered in the differential diagnosis of fever, even if the patient has been abroad.

Malaria

This is a tropical disease caused by *Plasmodium* spp.-*P. falciparum, P. vivax, P. ovale*, and *P. malariae* and transmitted by mosquitoes in Africa, Asia, and South America. The incubation period is from 7 days to several months.

Plasmodium falciparum **malaria** is a severe illness with headache, malaise, and myalgia, high fever with sweating, rigors, and vomiting. Complications include:
Cerebral malaria altered conscious level, seizures. Exclude hypoglycaemia and meningitis.
Haemolysis may cause severe anaemia, requiring transfusion. Haemoglobinaemia may result in acute renal failure.
Shock altered capillary permeability lead to result in

hypovolaemia and shock. This may be accompanied by multi-organ failure. Volume resuscitation with colloid should be commenced. Invasive haemodynamic monitoring in an ITU will be required.

Hypoglycaemia may occur prior to, or during treatment. The glucose sticks must be checked at presentation and regularly thereafter. Treatment of hypoglycaemia is with IV dextrose.

All patients with malaria due to *P. falciparum* require admission to a high-dependency unit. Those with complications, such as cerebral malaria, shock, or severe haemolysis, will require admission to an ITU.

Non-*falciparum* malaria This is due to *P. malariae*, *P. ovale*, and *P. vivax* has an acute illness similar to that caused by *P. falciparum*, with headache, myalgia and fever, but the severe complications do not occur and the mortality is low. Complications that may occur include splenic rupture, and, in chronic infection, anaemia and glomerulonephritis.

Diagnosis The diagnosis is made by visualizing malarial parasites on examination of thick and thin blood films in the haematology laboratory. If the initial samples are negative repeated films will need to be examined as the number of parasites in the blood varies.

Treatment
Supportive Attempt to lower the temperature by tepid sponging, fanning, and paracetamol. Once on the ward, careful fluid balance should be maintained and transfusion may be required. Monitor for signs of splenic rupture.

Specific antimalarial treatment Most patients can be treated with oral drugs. Indications for IV therapy are inability to take oral drugs and severe manifestations of malaria. The pattern of chloroquine resistance, and therefore recommendations for treatment, are changing. This book does not give any specific guidelines for treatment as they would soon become outdated. Current recommended treatment is obtainable in the *British National Formulary* or from your local infectious diseases unit.

Typhoid

This is caused by *Salmonella typhi* and transmitted by faecal–oral spread. The incubation period is 3–21 days. It occurs World-wide, but is uncommon in Northern Europe and North America.

Typhoid presents with malaise, headache, fever, cough, gastrointestinal disturbance, and abdominal pain. Confusion, meningism, and other CNS signs may occur. 'Rose spots' do not always occur. Patients should be admitted under the infectious diseases team. Treatment is with chloramphenicol or amoxycillin.

Yellow fever

This is caused by arboviruses and is transmitted by mosquitoes. The incubation period is 3–14 days. It occurs in Africa, the Southern United States, Central and South America, and the West Indies.

Yellow fever varies from a mild illness with fever, headache, and myalgia to a fulminating disease, resulting in death. In its severe form there is high fever, a bleeding tendency, jaundice, and renal failure. Treatment is supportive.

Lassa fever

This is caused by the Lassa virus and the incubation period is 3–16 days. It occurs in West Africa—Nigeria, Sierra Leone, Liberia.

Although Lassa fever is uncommon it is highly infectious. Health care personnel have died from Lassa fever and so strict barrier nursing and isolation must be instituted as soon as the diagnosis is suspected.

Its initial presentation is non-specific with fever, malaise, myalgia, headache, and sore throat. After 3–6 days the illness worsens with high fever, chest and abdominal pain, vomiting, and dehydration, pharyngitis, and facial oedema. Coma and shock usually precede cardiac arrest.

Management As soon as the diagnosis is suspected, strict isolation and barrier nursing must be instituted, and your own consultant and the local infectious diseases consultant should be contacted. Ideally the patient will be treated in an infectious diseases unit, but staff must be protected prior to and during transfer. Initial management is supportive.

Further reading

1. Oh, T. E. (1990). *Intensive care manual*. Butterworths, Sydney.
2. Schmidt, R. D. and Wolfe, R. E. (1992). Shock. In *Emergency medicine: concepts and clinical practice*, (ed. P. Rosen), pp. 163–72. Mosby-Year Books, St. Louis, MO.
3. Edwards, J. D. (1993). Management of septic shock. *British Medical Journal*, **306**, 1661–4.
4. Fisher, M. M. and Baldo, B. A. (1988). Acute anaphylactic reactions. *Medical Journal of Australia*, **149**, 34–8.
5. Lindzon, R. D. and Silvers, W. S. (1992). Allergy, hypersensitivity and anaphylaxis. In *Emergency medicine: concepts and clinical practice*, (ed. P. Rosen), pp. 1042–65. Mosby-Year Books, St. Louis, MO.
6. Pointer, J. E. (1992). Toxic shock syndrome. In *Emergency medicine: concepts and clinical practice*, (ed. P. Rosen), pp. 1970–1. Mosby-Year Books, St. Louis, MO.

CHAPTER 13

The elderly patient

Introduction

The greatest rate of decline in organ system function occurs during the fourth decade of life, but for most people, it is in the seventh and eighth decades that this relentless decline impacts on health and the activities of daily living. Elderly patients have very little reserve of cardiovascular, renal, hepatic, or neurological function and thus may decompensate following relatively trivial illness.

In addition to 'physiological' decline, organ systems are also affected by the increasing incidence of virtually all diseases with age. Cardiovascular and cerebrovascular disease, in particular, tend to limit function. This high incidence of disease results in a high incidence of medication, with attendant risk of adverse reactions and interactions.

The elderly in our society are also frequently socially isolated. The death of a spouse, isolation from children, and the responsibility of caring for oneself and one's home add to this burden.

Despite all of these factors, the majority of elderly people live in the community and are reasonably well. Never dismiss a problem as the product of ageing alone until you have sought a reversible cause, never assume that the elderly have poor quality of life. Never make assumptions based on chronological age: it is the individual patient you must assess and treat.

Manifestation of illness

The high incidence of multisystem disease and the lack of physiological reserve mean that elderly patients may present with decompensation in one system due to pathology in another. In particular, poor cerebrolvascular and cardiovascular function mean that patients may present with a confusional state or the picture of left ventricular failure secondary to respiratory tract or urinary tract disease. The other common presentations are incontinence, immobility, and falls. Changes may be more subtle with simply a 'failure to cope' in their usual environment without obvious manifestations of disease.

Social isolation and the lack of support means that elderly patients are at risk of prolonged illness before coming to medical care. Patients may present late in the course of illnesses which would otherwise be considered minor or, worse still, may lie in bed or on the floor at home with quite severe illness before being discovered.

It is vital that no elderly patient is discharged from the A&E department before a thorough history and examination, and at least a brief functional assessment. A thorough history includes corroboration from others if there is any doubt as to the patient's story, a full medication history, and a multisystem enquiry. A thorough examination includes all systems, lying and standing blood pressure, and dipstick urine examination. It is essential to stand the patient up and ensure that they can walk prior to discharge. Just as in younger patients, no elderly person should leave the A&E department alone unless they are able to walk out.

Finally, it is important to assess the underlying cause for any presentation. Fractures of the wrist, fractures of the neck of femur, and a variety of soft-tissue injuries following falls are common presentations in the elderly. It is not sufficient to say that the reason for the injury was a fall without attempting to determine the cause of the fall. Cardiac arrhythmias, postural hypotension, and transient ischaemic attacks must be identified and treated if a recurrence is to be avoided.

Be wary of patients presenting for 'social admission' or 'awaiting placement'. There is no such thing as a pure social admission, since all of these patients do have identified organic disease. In some cases, this may be chronic and without effective therapy, and the precipitating factor may be a problem related to the patient's carers. However, this precipitating problem must be carefully identified, and any component of medical disease sought and treated if possible.

Acute confusion

Assessment and management of the acutely confused elderly patient is little different from that of the younger patient (see Chapter 6). However, the incidence of confusion as a presenting condition is much higher in the elderly, and the spectrum of causes is a little different. In particular, urinary tract and respiratory tract infections which would not cause problems in a younger patient may cause a profound confusional state. In addition to inadequate physiological reserve, the elderly have limited response to illness, so that signs, symptoms, and investigations, such as white cell count, may not point towards the source of the problem.

As in other areas of emergency medicine, the priority is to identify and treat immediately reversible disease. Thus, confusion due to a small cerebral infarct may be a diagnosis of exclusion, confirmed on CT scanning at a later date. However, confusion due to urinary tract sepsis must be treated promptly.

By definition, confused patients do not give an adequate history of their illness. It is essential to obtain background history about the patient's usual mental state, degree of function, and the circumstances leading to their presentation. Always ask yourself the question: 'Why did this patient present now?' An elderly patient who normally lives alone in her own home and manages her own shopping, ironing, and cleaning must have acute, and therefore potentially treatable disease, if she is brought in by neighbours with a confusional state. A patient who presents apparently confused when his usual carer has become ill may have a dementia, with worsened confusion caused by removal from his usual environment. Both these patients require appropriate work-up, but the prognosis and expectation of findings in talking to the relatives will be very different.

Acute confusion in the elderly is a common presentation, and the basic investigational work-up is similar in most patients. This is an illness in which it is appropriate to take a series of blood tests, and order investigations before completing a thorough history and examination, so that the result may be more quickly available. Note that investigations are **not** a substitute for a thorough history and physical examination, but they are a necessary part of the search for occult disease that is prevalent in this patient group.

Immediate assessment and management

1. Airway, Breathing, Circulation.

2. Look for and treat reversible causes:

Hypoglycaemia Check the blood glucose by portable glucometer and give IV dextrose if necessary (see Chapter 7).

Hypoxia Check pulse oximetry. Give high-flow oxygen by facemask if there is any suspicion of respiratory problem.

Hypothermia Check core temperature. Low temperature is usually obvious on history and examination, and unfortunately common (see Chapter 10).

Narcotic intoxication This is very rare in the elderly population, but must always be considered.

Thiamine deficiency Just because a patient has reached his/her seventies does not mean that they are not an alcoholic. Give 100 mg of IV thiamine if suspected.

3. Establish intravenous access and take blood for fbc, U&E, creatinine, blood cultures if febrile, and ABG if there is any suspicion of hypoxia. (e.g. oxygen saturation <95 per cent on room air).

4. Start the process of obtaining ECG, chest X-ray, and MSU.

5. Undertake a thorough history and physical examination. As noted above, ensure that you have spoken to the relatives or carers and particularly sought a precipitating cause for this presentation.

Disposition

The patient will be admitted to hospital and the slow and often difficult task of preparing them for discharge needs to begin at the time of admission. The family, medical staff, and patient (when their acute confusional state is resolved) need to have a clear idea of the likely outcome of the hospital stay.

Incontinence

Incontinence, particularly urinary incontinence, is common in the elderly. It is usually a chronic problem and when it presents acutely, an underlying cause must be carefully sought. Specifically seek and exclude:

Urinary tract infection The most common cause in emergency medicine.

Urinary retention (due to prostatic enlargement in males) with overflow.

Constipation as a cause of faecal incontinence (impaction with overflow) and as an irritant causing urinary incontinence. Always perform a rectal examination.

Change in urine output commonly due to new diuretics or diabetes.

Immobility Patients (particularly those with some dementia) may wet their beds because it is now too difficult to get to the toilet. Look for causes of immobility (below) particularly occult injury such as fractured neck of femur.

Other causes Strokes, spinal cord lesions, pelvic floor prolapse.

Immobility

Elderly patients may present with immobility as a primary complaint, or, in the presence of dementia, may be brought by their carers with a 'complication' such as bed sores, incontinence, or poor hygiene. The cause may be found in any system, or may be multifactorial. Attempting to walk the patient with their usual walking aids is essential, and will usually identify the system responsible. Common acute causes to be excluded are:

'Falls' (see below) with fear of walking.

Cerebrovascular lesions particularly those affecting control of the lower limbs.

Remember that an uncommon midline cerebellar lesion will

cause great difficulty walking but may not affect finger–nose tests.

Occult injury, particularly fractured neck of femur and spinal crush fractures.

Confusional state often due to sepsis. Sometimes immobility may be the only symptom.

Social factors such as loss of walking aids or change in usual environment. A small number of patients may be attempting to manipulate their carers but do not leap to this conclusion.

Falls

Falls with or without significant injury are common in the elderly. Although often no cause is identified, it is essential to undertake a thorough search. In particular, look for:

Postural hypotension Take lying and standing blood pressure. This condition may be due to medication (especially antihypertensive or anti-anginal agents) or autonomic neuropathy.

Transient ischaemic attack Usually diagnosed on history alone, but check for carotid bruits.

Cardiac arrhythmias Always look at the ECG and if the history is suspicious, arrange follow-up with Holter monitoring.

Subtle neurological abnormality Carefully check for cerebellar signs, Parkinsonism, visual field defects, and minor lower limb weakness.

'Unable to cope'

All doctors who work in emergency medicine become familiar with the presentation of a patient who is no longer able to care for themselves, or to be cared for in the community. These patients are often labelled 'acopia', 'social problem', or 'placement problem', and most frequently present (sometimes with a referral letter) at times when other community services are not available. Labels such as these are unhelpful, since, as noted above, virtually all these patients have diagnosable medical illness, albeit frequently irreversible.

There are two major categories of such patients: those whose presentation is a marker of acute disease (falls, incontinence, confusion), and those whose disease is chronic and whose presentation has been precipitated by social factors (loss of carer, family pressure, domestic crisis). Unfortunately, these two groups cannot necessarily be distinguished at the time of presentation. It is essential to undertake a thorough history, examination, and investigation along the lines of that described for acute confusion above.

All such patients require admission to a health care facility while their problems are sorted out. Most commonly, this will be the hospital to which they present. If the patient has had regular contact with their GP, it is likely that other alternatives such as nursing homes, respite care, and domiciliary care will have been tried, excluded, or found to be unavailable. Despite the wealth of pathology that these patients offer to doctors and students alike, the need for admission is often resisted by medical units, on the grounds that these patients will occupy a bed for a long time, often without demonstrable improvement. **It must be stressed** that these patients do require admission and the hospital is frequently the last resort in their care.

Elder abuse

Much attention is focused on the very real problem of child abuse, but the frail elderly are another group without sufficient power to protect themselves. Abusers may be members of the elderly patient's family, paid carers, or staff at residential or nursing homes. As with injured children, seek an accurate history of the cause of any injury, looking for unusual, suspicious injuries, or injuries that do not fit the history given.

If you suspect a non-accidental injury contact a senior member of A&E department staff for advice and contact the social services. Under no circumstances must your patient be returned to an environment where they are at risk.

'Granny abuse' should be distinguished from 'granny dumping'. The latter is a phenomenon seen mainly in countries without free hospital care. Elderly, frail, and sometimes very ill patients are simply left in A&E departments by their

younger, fitter relatives. The motives are often purely finan-
cial, when the family do not wish to be identified because
they do not wish to pay for expensive medical care. However,
cases are reported where this method of presentation is used
by the family in desperation, in order to obtain respite care.

Discharging the elderly patient

Patients aged over 75 who present to A&E departments are a
generally ill group, with a 50 per cent admission rate and 10
per cent mortality at three months in most hospitals. Nevertheless, this means that half of them are sent home, perhaps
half of those alone to their own home. Clearly, it is essential to
ensure that they will cope and that they will receive appropriate care and follow up at home.

Doctors in A&E departments may be tempted to abbreviate
that part of the care involved in discharge. **Never leave out
steps in the discharge process.**

1. Check (again) that the patient is able to function in his/her
usual environment. Never send a patient home alone with
bilateral wrist plasters, crutches they cannot use, or a medication bottle they cannot open (all of these have happened), and
do not send them with a carer unless that carer is capable of
managing them.

2. Mobilize resources in the community to help, such as
meals-on-wheels, home help, bath assistant, district nurse, as
appropriate. Do not just assume the family will 'drop in'—tell
them to do so and when.

3. Ensure that the patient's GP is informed rather than relying
on the patient to call them. Preferably, write a letter *and* ring
the GP.

4. Ask the patient (or carers) to repeat the important information you have given them. If they do not understand their
treatment, medication, or follow-up, take the time to explain it
again. Not only does good discharge practice improve patient
care, it also improves outcome and hence A&E department
workload. Patients who are inadequately discharged are more
likely to return, often with a much more serious condition.
Preventative medicine works!

Further reading

1. Coakley, D. (1981). *Acute geriatric medicine*. Croom Helm, London.
2. McDonald, A. J. and Abrahams, S. T. (1990). Social emergencies in the elderly. *Emergency Medicine Clinics of North America*, **8**, 443–59.

CHAPTER 14

Transferring patients; and who should go to the ITU?

Interhospital transfer

Ill patients requiring specialist facilities that are unavailable locally may need to be transferred to another hospital. During transfer, patients are at risk, particularly from hypoxia and poor perfusion, so it is essential that every interhospital transfer is planned and carried out with meticulous attention to detail.

Emphasis must be placed on safe transfer *after* full stabilization, not speedy transfer of an unstable patient who may deteriorate *en route*. Any interventions that could be required prior to arrival at the receiving hospital must be performed prior to departure, as the simplest procedure becomes difficult or impossible in a moving, or even a stationary vehicle. Remember that ambulances can break down, get lost, or stuck in traffic so assume the journey will take at least twice as long as you expect.

Preparation of the patient

Airway   The airway must be open and protected during transfer. If the patient's airway is at risk because of present or potential impairment of conscious level or any other cause they will require anaesthesia and intubation. Tracheal tubes must be secured in position. Re-intubation may be impossible in transit.

Breathing All patients should have high concentration oxygen for the entire transfer to prevent hypoxia. If the patient has, or could develop, respiratory failure they should be ventilated with 100 per cent oxygen prior to transfer. Any pneumothorax, no matter how small, must be formally drained, using a drainage bag with a flutter valve, not an underwater seal which could break or spill.

Circulation The patient must have **two** wide-bore IV cannulae connected to fluid so that they can be flushed. They should be well secured in position and splinted if they are near a joint. If you transfer a patient with a single cannula it is guaranteed to come out or become blocked just when it is needed. If the patient is hypovolaemic they should be resuscitated before transfer. Monitoring of the ECG, blood pressure, and oxygen saturation is essential. If the patient is potentially unstable consider siting an arterial line for haemodynamic monitoring.

Temperature control Ambulances and hospital corridors are cold places and ill patients easily get hypothermia. After the patient is stable aim to cover their entire body, except the face, with both a metallized sheet and warm blankets. Ensure you can still get to a three-way tap on their IV lines and that you could get access to the chest if they should have a cardiac arrest.

Communication Tell the patient and their relatives what is happening and why the transfer is necessary. Ensure the relatives have time to see the patient before the transfer and know the name and whereabouts of the receiving hospital and ward.

Equipment

Every hospital should have transfer equipment prepared, checked and ready to go, but if this is not the case ABC is a useful reminder of what you need.

A Portable suction (check it works), rigid and flexible suction catheters, oral airways, tracheal tubes, including sizes above and below what you would expect to need, stylet, laryngoscope with spare batteries and bulb, and gum elastic bougie.

B Adequate oxygen (take at least twice as much as the maximum you expect to need), self-inflating bag with reservoir and appropriately sized anaesthetic masks and oxygen masks.

Pulse oximeter. If you think the patient might need to be ventilated *en route*, electively ventilate and connect to a transport ventilator prior to departure.

C Spare cannulae, tourniquets, tape, ECG monitor, blood pressure monitor, defibrillator. If the ambulance has a defibrillator check what sort it is as some semiautomatic defibrillators will not shock if there are pacing spikes. IV fluids and cross-matched or O-negative blood as appropriate.

Drugs Take an emergency drug box containing the drugs listed in Box 14.1

Box 14.1 Contents of the emergency drug box

- Adrenaline 1 mg ×3
- Atropine 1 mg × 3
- Lignocaine 100 mg × 2
- Diazepam 10 mg × 3
- Morphine 10 mg × 3
- Anti-emetic (e.g. metoclopramide 10 mg × 2)
- Anaesthetic drugs, selected by the transferring anaesthetist
- Specific drugs for the patient's condition

Remember that the patient may deteriorate so take more than you expect to need and second-line agents. If the patient is on an infusion this will need to be given through battery-driven syringe pumps, not burettes or drip counters, which do not function correctly in a moving vehicle.

Food, drink, and money Transfers can be long and it is unpleasant to end up in a strange, city, hungry, thirsty, and with no means of getting home. Take sandwiches and hot drinks with you for the journey and enough money to get home.

Monitoring

During transfers many clinical signs become hard to elicit. Colour is changed by the tinted windows and street lights, the pulse cannot be felt because of the vibration and heart sounds, breath sounds, and blood pressure are drowned by engine noise and sirens. Monitoring is essential, at the very least, ECG monitoring, pulse oximetry, and blood pressure monitoring. Transport monitors should be lightweight and sturdy, have a long battery life, not suffer from motion artefact, and, most importantly, be thoroughly understood by all the transfer team.

Who should transfer the patient?

In an ideal world every hospital would have a transfer team of senior doctors and nurses, experienced in managing critically ill patients in transfer, with dedicated equipment. This is not always the case. If your hospital cannot provide suitable staff and equipment contact your consultant and consider asking the receiving hospital to send a retrieval team. No patient should be transferred in a less than optimal manner.

The staff transferring an ill patient must have the skill and seniority to deal with every problem that could occur so the most senior available personnel should make up the team. If the patient has actual or potential problems with their airway or ventilation the doctor must have anaesthetic skills, being happy to manage a paralysed patient on a ventilator. This is likely to be a registrar, or above, in anaesthetics or ITU, but may be a senior physician or A&E doctor. Paramedics may offer to take the place of a nurse on your team, and for more straightforward transfers this may work well but if the patient has invasive monitoring and multiple infusions an A&E or ITU nurse will be required.

Aerial transfers

Everything that applies to road transfers applies to helicopter and fixed-wing aerial transfers but more so. The medical staff in the team must be familiar with safety procedures in and around aircraft and the physiological effects of reduced pressure at altitude. The patient must be protected against the cold and have their core temperature monitored.

The most you can achieve in-flight is to inject a drug into an pre-existing IV line or alter the settings on an infusion pump, so stabilization prior to transfer is essential. If there is any possibility of the patient deteriorating they must be electively ventilated prior to departure as IV cannulation and intubation are impossible in an aircraft. Because of the lower atmospheric pressure all patients need high concentration oxygen and the patient should have a naso- or orogastric tube passed to decompress the stomach. Even the smallest pneumothorax must be drained to prevent a tension pneumothorax developing.

Intrahospital transfer

The importance of safe controlled interhospital transfer is being increasingly recognized but it is important to remember that the journey from the A&E department to CCU or ITU can be complicated by hypoxia and haemodynamic deterioration. The doctor who organizes the safest transfers is a pessimist, and when deciding on the staff and equipment to accompany an ill patient assumes that the lift will get stuck and the patient will deteriorate.

As an example, on transferring a patient with an acute myocardial infarction from the A&E department to the CCU they should be on high-flow oxygen from a full cylinder, with secure IV access, accompanied by full resuscitation equipment, including airway adjuncts, self-inflating bag with reservoir and mask, intubation equipment, resuscitation drugs and connected to a monitor-defibrillator, together with a member of staff proficient in resuscitation. Similarly, if a sick asthmatic is being transferred to the ITU for ventilation they should be anaesthetized and ventilated prior to leaving the resuscitation room.

Who should go to the ITU?

Any patient, regardless of age, who would benefit from admission to an ITU should be referred for admission. Patients with acute respiratory or circulatory failure obviously need ITU referral but patients at risk of these conditions should also be referred so that if deterioration does occur the necessary facilities and expertise are immediately available. The corridor between the medical ward and the ITU is a dangerous place for an ill patient.

Airway problems

A decreased level of consciousness leading to a risk of obstruction or aspiration is the most common reason for medical patients requiring ITU admission. Examples include patients following overdose and cardiac arrest. Any patient who does not have a normal gag and cough reflex requires airway protection with a cuffed endotrachael tube. Many of these patients will require ventilatory support. Rarely, airway obstruction may occur due to infections, such as epiglottitis, and these patients must be managed by senior anaesthetic staff.

Breathing problems

'Medical' causes for respiratory failure requiring ITU admission include:

- Gas exchange problems (e.g. pneumonia, pulmonary emboli, ARDS).
- Bronchospasm.
- Inadequate or absent respiratory drive (e.g. overdose, intoxication, post-cardiac arrest) this often co-exists with a need for airway protection.
- Muscular problems (e.g. exhaustion from prolonged respiratory distress, Guillain–Barré syndrome).

The patient's clinical state is as important as absolute P_{O_2} and P_{CO_2} levels. Patients should not be allowed to become exhausted, hypoxic, and acidotic before ventilation but the approach of respiratory failure should be anticipated and pre-empted. Established respiratory acidosis, CO_2 retention, and hypoxia despite oxygen therapy, are all indications for urgent ITU referral. Assessment of ill patients must include arterial blood gas analysis, as pulse oximetry can be normal despite severe acidosis and hypercapnia.

Circulatory failure

Shock may be due to hypovolaemia, sepsis, anaphylaxis, or cardiac failure. Apart from hypovolaemic shock due to haemorrhage where immediate surgery may be required, shock persisting after initial resuscitation is an indication for ITU management for invasive monitoring. Shock is assessed not only by blood pressure and heart rate, which change late in the development of shock, but also by peripheral perfusion, mental state, and urine output. The presence of a metabolic acidosis on blood gas analysis is a good indication of cellular hypoxia, which, if not rapidly reversed by resuscitation, is an indication for ITU referral.

Patient transfer

As outlined on p. 194 intrahospital transfers can be dangerous leading to hypoxia and hypotension. Great care must be taken to stabilize the patient prior to leaving the safety of the resuscitation room. Occasionally, there is a problem contacting the in-patient teams about critically ill patients. If this occurs, after discussion with your own seniors, you must persevere, going higher up the medical hierarchy, until you contact a member of the team. It may make you temporarily unpopular with your colleague who did not answer his/her bleep but the patient's needs must come first.

'Not for the ITU'

There are some patients whose severe illness is the terminal event of an incurable illness, where aggressive treatment will not be able to restore a quality of life that the patient would find worthwhile. This is well accepted in the setting of disseminated malignancy and dementia but can also be the case in other conditions, such as chronic respiratory failure due to chronic airway obstruction, severe heart failure in patients not suitable for transplantation, and end-stage HIV-related disease.

If a patient has no possibility of recovery it is not appropriate to subject them to invasive procedures. It should be clearly marked in the notes that the patient would not benefit from ITU treatment and is not for resuscitation in the event of a cardiac arrest and they should be kept pain-free and undistressed to die with dignity with their family around them.

Further reading

1. Oh, T. E. (1990). *Intensive care manual.* Butterworths, Sydney.
2. McLellan, B. A., Fulton, L. A. *et al.*, (1988). Long distance critical care transport. *Emergency Prehospital Medicine*, **2**, 33–9.

CHAPTER 15

Breaking bad news

Introduction

During the treatment of every critically ill patient a member of the resuscitation team should keep the relatives informed, as even a short resuscitation can seem endless to relatives left without any information. The relatives should be able to be together, away from the public gaze, preferably in a designated relatives room (Box 15.1).

If there is not a relatives room then Sister's office will prob-

---

Box 15.1 **The ideal relatives room**

- Private and quiet
- Comfortable chairs/settees
- Telephone with outside line
- Tea and coffee making facilities
- Smoking permitted and ashtrays provided
- Easy access to toilets

---

ably be the most appropriate place. No one should be told bad news in a corridor or busy waiting room.

Who should tell the family?

This is usually the job of the resuscitation team leader. If another team member has already been keeping the family

informed of progress they may wish to tell them the bad news.
The most senior doctor should always talk to the relatives but
it is a good idea to be accompanied by the team member who
has already seen them.

Preparation

It is worth spending a moment preparing yourself to speak to
the family.

Physical preparation
1. Wash all traces of blood, betadine, etc, from your face and
   hands.
2. Take a critical look at your white coat. If it looks as if you
   work in an abbatoir change it.

Mental preparation
1. Check the patient's name, and the relatives' names and
   relationship to the patient.
2. Check where the relatives are. There may be more than one
   set of relatives and you do not want to tell the wrong ones
   the bad news.

What should be said

There is no magic formula that is right for every relative or
every doctor. You must say what you feel happy with. The
following paragraphs contain a few ideas.

Introduce yourself as *Doctor* X.   If you introduce yourself by
your name alone they may not realize you are a doctor. Similarly, if you are usually known as Mr, Mrs, Miss, or Ms, say
you are a doctor, otherwise they may think the doctor did not
speak to them.

Tell the truth simply   Avoid medical jargon. 'Ventricular fibrillation' means nothing to most people whereas 'his heart has
stopped beating' can be easily understood. It is always difficult to tell sad news but do not be tempted to raise false hopes
to make your job easier.

If the patient has died say so It is important to say clearly that the patient **has died,** or **is dead**. Phrases, such as 'passed on', 'gone to a better place', or 'we have lost him', can be easily misunderstood, particularly by relatives desperate for good news.

Blame and guilt It is common for relatives to ask 'Is there anything I could have done?' There is only one answer to this question, which is 'No'. (This is not the time to preach community cardiopulmonary resuscitation.) Reassuring words from you about how they are not to blame may prevent guilt later. Anger is a part of the bereavement process. Do not take it personally. Let the family know that you are there if they have any additional questions in the coming hour or weeks.

Religion

It is important to respect the religious beliefs of the family, even if you find them difficult to understand. Equally, if you yourself have strong religious beliefs you cannot assume that the relatives share them. Offer to call a minister of religion, as in their distress, the family may forget to ask. Many religions have regulations about how the body should be treated. Every department should have a copy of these and they should be followed.

Seeing the body

The relatives should be offered the opportunity to see the body, as most will want to, to say goodbye, but they may not know it is allowed unless you offer. It is common for friends to attempt to speak for a bereaved spouse thinking that seeing the body would upset them but the decision should be made by each individual. The relatives should be accompanied, at least initially, by a nurse who may need to 'give permission' for them to touch or hold the body. People need a variable length of time to say their goodbyes and they should not be hurried even if the department is busy.

Team debriefing

Ideally, each resuscitation should be followed by a debriefing. In a busy hospital this may not always be possible but the team leader's responsibilities extend beyond the patient to the team members. A 'thank-you' to acknowledge everyone's hard work is essential every time and an assessment should be made of whether anything more is required.

Training needs

If there were major problems in the team during the resuscitation, the period immediately afterwards when everyone is wound up is unlikely to be the most effective time to deal with them. A note should be made and training instituted as a result.

Emotional needs

Many people working in acute medicine carry with them memories of resuscitations which they left distressed and alone without anyone to talk to. One of the most important roles of the team leader is identifying and where possible alleviating this distress in team members and student observers. This is best done informally talking through the resuscitation, with the more senior staff admitting their own difficulties so that no one feels inadequate for being upset.

What about me?

As team leader your perceived failure may be hard to accept. Being a support for the other team members and observing the distress of the bereaved family may make things worse, particularly if they are hostile towards you. You too need to talk about the resuscitation and your feelings about it to someone who understands. This is likely to be someone in the same or a similar specialty, either medical or nursing. It is not a failing to be upset, it is a sign we are all still human.

Further reading

1. Handley, A. J. and Swain, A. (1994). *Advanced life support manual.* Resuscitation Council (UK).
2. Fallowfield, L. (1993). Giving sad and bad news. *Lancet*, **341**, 476–8.
3. Finlay, I. and Dallimore, D. (1991). Your child is dead. *British Medical Journal*, **302**, 1524–5.

Practical procedures

Introduction

The chapter is meant to provide a theoretical basis for those learning practical procedures under supervision, or to act as a reminder to those who have received such training. It is not supposed to be a do-it-yourself guide for beginners. The only safe way to learn a procedure is by observing it and then performing it under the supervision of someone skilled and experienced in that technique. Never agree to undertake a procedure on a patient unless you have been trained in that procedure and have achieved competence. It is better to call your boss in at 3 a.m. than harm the patient—it is what the patient would want.

Tracheal intubation

This keeps the airway open and protected but it is a technique that must be learnt under supervision and practised if competence is to be maintained. Intubation should be learned on manikins or in the controlled environment of a theatre anaesthetic room, not on a critically ill patient.

Almost all patients' airways can be maintained open using the simple techniques of head-tilt and chin-lift or jaw-thrust, possibly augmented by an oropharyngeal airway. If it is decided that tracheal intubation is required, to achieve airway protection, the most important thing is to maintain oxygenation. It is not a fault to fail to intubate, some patients are extremely difficult to intubate, but it is a fault to fail to maintain oxygenation during intubation.

Preparation

Pre-oxygenation The most important part of preparation in ensuring that the patient is pre-oxygenated, having received high concentration oxygen through a well-fitting mask.

Position Ensure the patient is in the optimal position, the 'sniffing the morning air' position with the neck flexed and the head extended on the neck. This makes visualization of the cords much easier.

Monitoring Ensure that the patient is connected to a cardiac monitor and a pulse oximeter.

Necessary equipment

- Functioning suction, with both rigid (Yankauer) and flexible catheters.
- A self-inflating bag with anaesthetic mask and reservoir bag, connected to oxygen.
- Appropriately sized cuffed tracheal tubes, cut to length, with the cuffs checked for leaks. The correct size is usually 8 for women and 9 for men. Have a half size larger and smaller available.
- Consider adding rigidity to the tube by using an introducer if you have been trained in their use.
- Laryngoscope. Check it works and have a spare available in case it stops working.
- Syringe to inflate the cuff, tape to secure the tube in place, and a stethoscope.

Technique

Take a deep breath and hold it. When you need to take another breath the patient does too.

Hold the laryngoscope in your left hand. Pass it into the right-hand side of the mouth, moving the tongue to the left. The jaw and tongue are lifted up as the laryngoscope is slid into the vallecula (see Fig. 16.1). When you can see the cords

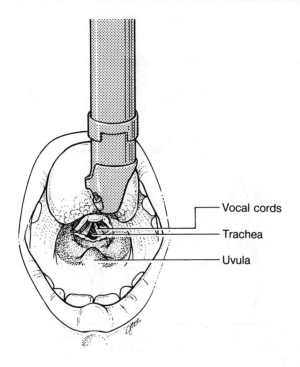

Vocal cords

Trachea

Uvula

Fig 16.1 • Advanced airway management 1: laryngoscopic examination

insert the tube, held in your right hand, so that it does not obstruct your view and pass it between the cords under direct vision (see Fig. 16.2).

Once you have seen the tube pass between the cords, inflate the cuff until there is no air leak audible during ventilation, and check its position. Auscultate both sides of the chest in the midaxillary line, and over the epigastrium. If it is correctly placed secure it in position with a tie or adhesive tape.

Ensure the patient is being ventilated with high concentration oxygen. If at any point you have difficulty, or feel you need to take a breath yourself before the tube is in place, remove the laryngoscope and return to oxygenating the patient with the bag-valve-mask.

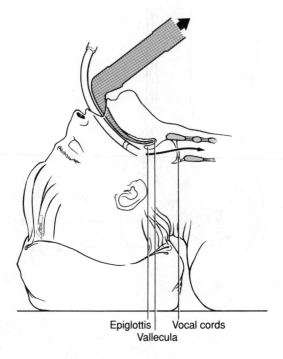

Epiglottis | Vocal cords
Vallecula

Fig 16.2 • Advanced airway management 2: oral endotracheal intubation

Order a chest X-ray to check the end of the tube is in the trachea and check the arterial blood gases. Continue to monitor the ECG, oxygen saturation, blood pressure, and, if available, end-tidal CO_2. A fall in the oxygen saturation suggests tube misplacement or inadequate ventilation. If you think the tube could be in the wrong place, remove it, re-oxygenate with a bag-valve-mask and attempt re-intubation when more senior help is available.

Cricothyroidotomy

The cricothyroid membrane is the safest and simplest place to obtain surgical airway access in an emergency. It is easily palpable, avascular, and a safe distance from the thyroid gland (superficial and below) and vocal cords (deep and above).

Cricothyroidotomy is indicated in supralaryngeal airway obstruction when tracheal intubation is not possible, such as in epiglottitis, burns, or facial trauma. It may rarely be needed for ventilatory support when the patient's anatomy renders tracheal intubation and ventilation impossible or impractical.

By definition, this procedure will be undertaken for the first time as an emergency. It pays to be familiar with equipment available in your A&E department for this purpose. Do not allow a patient to die of airway obstruction because you are uncertain of this procedure.

Technique

1. Lie the patient supine with the neck extended, if it is safe (i.e. take care in cases of trauma with possible cervical spine injury).

2. Identify the cricothyroid membrane which is a horizontal depression in the midline anteriorly just below the notch of the thyroid cartilage and just above the cricoid cartilage.

3. Prepare the skin with antiseptic and, if there is time, inject local anaesthetic.

4. There are prepackaged kits containing a guarded scalpel, cricothyroidotomy tube with stylet, and appropriate ties and connectors (e.g. Minitrach), alternatively, a standard tracheostomy tube or tracheal tube can be used. Make a short (1.5 cm) horizontal incision across the cricothyroid membrane, then cut the skin and membrane with the scalpel, to a depth of 1 cm. Twist the scalpel in order to open a gap between the thyroid and cricoid cartilages. This sounds simple, and looks straightforward on videos using cadavers, but in reality, the patient will buck and cough, and breathe violently through the hole created. The operator and assistant may be showered with blood.

5. Insert the tube through the hole in a downward direction, remove the trochar and apply ventilation or oxygen therapy as appropriate. Secure the tube.

Note that the uncuffed tube available in most kits, such as the Minitrach, provides airway access but **not** airway protection. If the cricothyroidotomy is to be used for more than one hour, or if there is any risk of aspiration, then you must use a

5 or 6 mm cuffed tracheal tube. This requires a 2.5 cm skin incision and a 2 cm cricothyroid incision.

Alternative incision technique If the cricothyroid membrane is not easily palpable, use a vertical incision of 2.5 cm in the midline, starting just below the thyroid notch. This has the advantage of being less vascular, but an assistant is then required to spread the edges of the wound, so that the cricothyroid membrane can be located.

Needle cricothyroidotomy The simplest approach to cricothyroidotomy is to use a wide-bore (12 or 14 gauge) IV cannula attached to a syringe. When air is aspirated, the tip of the needle is in the trachea. Adult patients cannot maintain adequate ventilation through this orifice, but it can be used for 'trans-tracheal jet ventilation'. Attach a high-flow (15 l/min) oxygen line to a Y connector and the cannula and occlude the free end of the Y for one second in every five. This method will cause hypercapnia, but with normal lung function, will maintain oxygenation for about 30 minutes.

Chest drain insertion

The method of chest drain insertion that is described below is safe in medical and trauma patients and is well suited to a setting where the patient's previous history is unknown.

Indications The usual medical indications for insertion of a chest drain in A&E are a large simple pneumothorax, which has not responded to aspiration, or less commonly, a large pleural effusion, haemothorax, or empyema.

A chest drain is not the emergency treatment for a tension pneumothorax—the patient would be dead by the time it was inserted. Convert the tension pneumothorax to a simple pneumothorax by placing a wide-bore cannula (e.g. 14G) in the 2nd intercostal space midclavicular line, to decompress the tension, then put in the chest drain.

Site The preferred site is the 4th or 5th intercostal space in the midaxillary line. The 2nd intercostal space midclavicular line results in obvious scarring and is more difficult and painful because of the greater muscle bulk.

Size In an emergency situation, a large drain should be

inserted so that it does not block. A size 28 Ch would be suitable. If there is blood or pus in the chest use the largest size available. In an average adult this will be size 32–36 Ch, but will depend on the size of the patient's intercostal spaces.

Technique

Every intrathoracic, and many intraabdominal organs have been speared by chest drain trocars and patients have died as a result. Do not use the trocar.

- Explain to the patient what you are going to do. Position the patient, find and mark the correct position. It is easy to put the drain dangerously low unless you count the rib spaces and mark the skin.
- Clean the skin and drape. Use strict aseptic technique.
- Infiltrate the skin and down to pleura with lignocaine (maximum dose 3 mg/kg).
- Make a 3–4 cm incision just above and parallel to the top of the rib (thus avoiding the neurovascular bundle), through the skin down to fat, then put down the scalpel.
- Using blunt dissection with a large pair of forceps (such as Spencer Wells forceps) divide the muscles down to pleura and make a hole in the pleura using the tip of the forceps.
- Put your gloved finger through the hole to check you are in the pleural space.
- Clamp the forceps on to the chest drain, the end of the forceps protruding slightly beyond the end of the drain.
- Push the drain and forceps into the pleural space, removing the forceps and advancing the drain once it has passed between the ribs.
- Connect the drain to an underwater seal system or a drainage bag/valve. If using an underwater seal, check the level is swinging, or bubbling.
- Secure the drain in position. Use a thick silk suture (0 or number 1 thickness) to secure the drain to the skin. Close the rest of the incision with thinner suture material (e.g. 2/0 silk). Tape the chest drain in for extra security.

Post-chest drain management

Get a chest X-ray to check the position of the drain.

The chest drain should not be clamped until it has finally

stopped draining or bubbling, as putting on a clamp could make a simple pneumothorax tension. The patient should be transferred to the ward with a nurse, who has instructions to keep the drainage bottle below chest level but not to clamp the tubing.

Venous access techniques

Intravenous access is essential in many emergency situations. A variety of access devices are available, and it is very important to be familiar with both the device and the technique that you intend to use. If you must obtain access urgently and the usual sites (forearms, cubital fossa, long saphenous vein) are unavailable or impossible, forcing you to use an unfamiliar site, such as the femoral or external jugular veins, do not use this opportunity to try out your skills with an unfamiliar technique (e.g. Seldinger). If you are only familiar with conventional IV catheters and have failed to obtain access peripherally, use the femoral veins or a venous cut down to saphenous or cubital veins. These are the safest sites if something goes wrong. There are four major techniques for IV access from the skin:

1. **Catheter over a needle** (e.g. Venflon). This is the familiar IV line used in most hospitals. It is simple to use but of limited length and diameter.

2. **Catheter over a needle with dilator/guide wire** (e.g. Arrow Emergency Infusion Device). There are a variety of these systems, designed to allow placement of short, wide-bore catheters in superficial veins.

3. **Catheter within a needle** These systems allow placement of long but narrow-bore intravenous catheters into deep or central veins.

4. **Seldinger technique** Prepackaged kits are available of single, double, or triple lumen central lines. There are a variety of sizes, but again the maximum diameter is limited, usually to about 14 gauge. Briefly, a needle is inserted into the vessel, a wire is passed through the needle which is then removed, and a dilator and then a catheter is passed over the wire. This technique requires a sterile field, cardiac monitoring, if there

is any possibility of the wire reaching the heart, and experience with the equipment. It is not a technique to try for the first time on a critically ill patient.

As a general rule, intravenous access for fluid resuscitation needs to be wide bore so the cubital veins, femoral veins, saphenus veins, and external jugular veins should be approached in roughly that order, according to availability. Central venous access (i.e. access to the superior vena cava) on the other hand, is usually performed for injection of drugs, such as inotropes, and for the monitoring of central venous pressure. Thus, smaller lines can be used. If this is the indication for inserting a central line, then it is appropriate to use the technique favoured by your ITU. The line will remain *in situ* for some time and will require to be maintained on the wards, so the site, dressing, and number of lumens should reflect the techniques used elsewhere in your hospital.

The Seldinger technique

This is not a DIY guide for beginners; it is a reminder for someone who has been taught the technique under supervision.

1. Assemble the equipment required: sterile tray with drapes, preparation swabs, skin disinfectant, local anaesthetic with syringe and needle, scalpel, needle holders, scissors, and appropriate suture (e.g. 2/0 silk), plus a prepackaged central line 'kit' with syringe, needle, guidewire, dilator, and central line. In addition, if you will be taking blood or injecting rather than connecting to an IV infusion, collect appropriate syringes and draw up solutions now.

2. Explain to the patient what you are going to do, and why, and obtain informed consent.

3. Position the patient appropriately (see p. 213), and ensure that any necessary resuscitation and monitoring continue. In particular, for jugular and subclavian lines, connect the patient to an ECG monitor which the operator can see.

4. Use a sterile technique with gown, gloves, mask, and adequate skin preparation.

5. Inject local anaesthetic (e.g. 5 ml 1 per cent lignocaine) into the skin and down the track of the catheter, taking care to avoid intravascular injection.

6. Using the long wide-bore needle attached to the syringe from the kit, puncture the skin and seek the vessel whilst applying gentle suction to the syringe. When you hit the vessel there should be free flush-back of blood. Advance the needle another 2 mm so that the tip is completely in the vessel, ensuring that flush back continues.

7. Pass the guidewire through the needle so that 5–10 cm is in the vessel (i.e. to the 10–15 cm mark at the needle hub). In most cases, this means removing the syringe and passing the wire directly into the needle, but some kits allow the wire to be passed down the barrel of the syringe. If the monitor shows ventricular ectopic beats or ventricular tachycardia the wire is in the heart and must immediately be withdrawn until the arrhythmia stops.

8. Remove the needle (and syringe if still attached) without disturbing the position of the wire.

9. Incise the skin at the entry site of the wire, making a hold large enough for the catheter.

10. Pass the dilator down the wire and into the vessel (again taking care not to disturb the wire position) then remove the dilator, leaving the wire in position.

11. Pass the catheter over the wire and into the vessel. **Never lose control of the wire**. The distal end of the wire must be grasped where it exits the catheter before the proximal end is covered by the catheter tip. If you cannot do this then withdraw the wire from the patient until you can.

12. Remove the wire, check for free blood flow-back through each lumen, then flush the catheter or connect to appropriate IV infusions.

13. Secure the catheter with sutures and apply a dressing.

14. Obtain a Chest X-ray for jugular or subclavian catheters to check for tip position and pneumothorax.

Subclavian line

The subclavian vein lies below and behind the clavicle, in front of the subclavian artery. It passes over the first rib and lies above, and in front of, the apex of the lung before passing downwards to join the great vessels. These relations give the infraclavicular approach to the subclavian vein the highest complication rate (up to 5 per cent) on insertion, but the flat skin below the clavicle makes it a very easy line to maintain, and therefore is the most popular in many ITU settings.

Infraclavicular approach
Position In a patient with lung pathology choose the affected side, so that a pneumothorax does not affect the only 'good' lung. Place the patient in 10 degrees head-down, turn the head slightly away from the side of insertion, and have an assistant gently pull the adducted arm towards the feet.
Entry Puncture the skin just below the midpoint of the clavicle and aim with the needle parallel to the floor for the sternal notch. It may be difficult to negotiate between the clavicle and the first rib.
Technique Once in the vein, turn the needle so the bevel faces towards the patient's feet before inserting the wire.
Problems Pneumothorax (more likely with a more medial approach), piercing subclavian artery or cephalic vein (more likely with a more lateral approach), passage of catheter into either jugular vein (always perform a check X-ray). If you hit the artery, withdraw and apply firm pressure for a full 5 minutes. It may require several passes to hit the vein, and if you fail then a more medial or more lateral approach is reasonable.

Supraclavicular approach
Position Head-down 10 degrees, turn head away from side of entry. Identify the clavicle and the sternal head of sternocliedomastoid.
Entry 1 cm lateral to sternocleidomastoid tendon and 1 cm above the clavicle. Aim at the contralateral nipple. The subclavian vein is quite superficial here: 1–2 cm only.
Technique Turn bevel downwards after entering the vein.
Problems As for supraclavicular approach (always perform a check chest X-ray). However, the entry point in the supraclavicular fossa is more difficult to care for.

Internal jugular line

This vein offers the straightest, most direct route to the superior vena cava. However, it is the site most susceptible to displacement by neck movement.

Position Head-down 10 degrees, head turned slightly away from the site of injury (turning too far moves the carotid artery forward). Identify the two heads of sternocleidomastoid and the pulse of the internal carotid artery.
Entry Pierce the skin just above the point of the triangle

formed by the two heads of sternocleidomastoid, aiming parallel to, and 1–1.5 cm lateral to the internal carotid artery.

Technique There are many vital structures in the neck: take care not to probe around blindly. If the approach fails, try more laterally rather than medially.

Problems Carotid artery puncture can occur. Pneumothorax is less common than with subclavian line insertion. Always perform a check chest X-ray.

External jugular line

This vein crosses the posterior border of sternocleidomastoid at the level of the C6 transverse process and runs forward and down along the sternocleidomastoid before joining the internal jugular. If venous pressure is raised, it is usually clearly visible. The vein may, however, follow a tortuous route, and contain significant valves. Thus, this readily available and safe site has a low incidence of complications, but a relatively high incidence of failure to cannulate with long central venous pressure lines.

Position Head-down 10 degrees, head turned slightly away from the site of entry. Identify and palpate the vein directly.

Entry Enter it as you would any large superficial vein, preferably in a straight section.

Technique Use gentle rotation and repositioning of the wire or catheter if there is a problem with tortuosity or valves.

Problems Apart from failure to cannulate, local bleeding and venous perforation which can be treated by local pressure.

Femoral venous line

The femoral vein is large and superficial, but the groin is a particularly inconvenient site for all but bed-bound patients. Where wide-bore venous access is required in an emergency, this is probably the easiest site, but it should be avoided in the presence of abdominal pathology. A tense abdomen, particularly from bleeding, may significantly obstruct inferior vena caval flow.

Position Supine, hips in neutral position. Identify the femoral arterial pulse and the inguinal ligament.

Entry 1 cm medial to the pulsation of femoral artery just below the inguinal ligament. Aim at 30 degrees to the horizontal in a line parallel to the patient's spine.

Technique Gentle probing is acceptable, but try more medially rather than laterally, in order to avoid the artery.

Problems Arterial puncture and local haematoma.

Venous cut down

This is an emergency technique to be used when other methods are contraindicated or staff do not have the required skills. The two possible sites are the long saphenous vein just anterior to the medial malleolus, or the cephalic vein in the lateral part of the cubital fossa (remember that the medial part of the cubital fossa contains the brachial artery).

1. Assemble equipment required: sterile tray or dressing pack, scalpel, artery forceps, cannula (special cannulae are available, but any wide-bore cannula will do).
2. Explain to the patient what you are going to do, and why, and if possible obtain informed consent. Before incising an ankle, always check for scars of varicose vein surgery or venous harvesting for arterial grafts.
3. Use a sterile technique.
4. Make a 3 cm transverse incision over the site of the vein.
5. Dissect out the vein using fine artery forceps.
6. Lift the vein with artery forceps behind it and insert the cannula. Use a conventional IV cannula, or make a small nick in the vein with a scalpel and insert a larger cannula.
7. Secure the cannula with sutures and close the skin. Apply pressure to the site. It is not necessary to tie off the distal part of the vein if the cannula is a good fit in the lumen. This is a temporary technique, which can be used for 1–2 days, but is best replaced by another form of access as soon as expert help is available.

Removing a central line

Removal of a central line is not commonly indicated in the A&E department, but is sometimes required when a line is blocked, infected, or misplaced (e.g. tip of a subclavian line high in the jugular vein). Lie the patient down, have an airtight dressing ready, and cut any sutures. Pull the line out smoothly, and immediately apply pressure over the site with the airtight dressing. If the line has been present for some time then there is a small but real risk of air embolism if the site is not covered. Keep the tip sterile if it is to be sent for culture.

Pacing

See Chapter 3 for the indications for pacing, and the resuscitation that should precede the pacing and continue while it is being undertaken.

External pacing

External pacing, like an isoprenaline infusion, is only a holding measure until transvenous pacing can be performed. It is extremely effective but can only be used for a limited period because of patient discomfort, so as soon as you feel pacing is required, ensure someone in your team is setting up for the insertion of a pacing wire and calling for senior help.

The controls of every machine vary, so check the one in your department before you have to use it. In external pacing, a current is passed between two large diameter adhesive pads, which are applied, one on the anterior and one on the posterior surfaces of the chest. If the patient is conscious warn them that they are likely to feel some discomfort and turn it on, in 'demand' mode, at a rate of about 80/minute. Adjust the output until you find the pacing threshold (i.e. the level that just captures the ventricles), and set it slightly above this level. Some patients find the discomfort from the inevitable pacing of the pectoral muscles intolerable and require analgesia. Since the development of modern external pacing machines the need for oesophageal pacing has diminished.

Temporary transvenous pacing

In this technique, a pacing wire is passed through a vein into the right ventricle. The wire is then connected to a pacing box and achieve ventricular capture. This technique must be learned under supervision. Do not perform this procedure unless you are familiar with the technique and the equipment you will be using.

Necessary preparation

Full resuscitation equipment, including a defibrillator, and adequate, experienced staff.

Correct equipment, including introducer, wire, pacing box with fresh batteries.

Image intensifier. It is possible to insert a pacing wire 'blind', but it greatly reduces the chance of ventricular capture so it is not recommended.

Technique

Strict aseptic technique is essential.

The introducer is sited in a large vein (the subclavian, internal jugular, femoral, or cephalic veins are suitable) see pp. 212–15

The wire is passed through the introducer and advanced under X-ray control, using the image intensifier. It is passed from the right atrium, through the tricuspid valve into the right ventricle.

As the wire crosses the tricuspid valve it is seen to cross the midline on the image intensifier screen and there may be a run of ventricular extrasystoles. Occasionally, the wire triggers ventricular tachycardia or ventricular fibrillation, thus one team member must constantly watch the monitor. Once the wire appears to be at the apex of the right ventricle the pacing box should be connected and turned on. Keep it on the 'demand' setting so that pacing impulses are only produced when the heart rate falls below a set rate, say 80/minute. Ensure that the patient's own ventricular complexes are being sensed. Establish the capture threshold, that level of pacemaker output at which the ventricle is just captured, this should ideally be at or below 1 volt. Set the output to 3 volts, well above the level that captures the ventricle.

Any patient with a temporary pacing wire must be on a CCU or ITU and continuously monitored.

Emergency pericardiocentesis

This section is aimed at the resuscitation team leader who, having diagnosed cardiac tamponade, has to perform emergency pericardiocentesis (e.g. in the presence of electromechanical dissociation cardiac arrest, or severe hypotension when cardiac arrest is felt to be imminent). If pericardiocentesis is performed in a non-emergency situation, such as in a patient with slight breathlessness but reasonable blood pressure and perfusion, it should be carried out under ultrasound control by a cardiologist or another senior clinician experienced in the procedure.

Pericardiocentesis is a sterile procedure done under aseptic technique.

Resuscitation At all times, ensure the patient is safe with ABC maintained. Do not get so engrossed in the procedure you forget the patient and do not notice a deterioration.

Site Pericardiocentesis through the 4th intercostal space left of the sternum is not recommended as it frequently results in coronary artery damage. The preferred site in all situations is to insert the needle 2–3 cm below the xiphisternum.

Patient position If the patient has suffered a cardiac arrest they will be prone with cardiopulmonary resuscitation in progress. If the patient still has a cardiac output they should be sat up against pillows at 45 degrees.

Monitoring In a non-emergency situation this procedure will be done under ultrasound and ECG control. In a coronary care unit there will be specialized pericardiocentesis kits that enable the needle to be connected to the V1 lead of an ECG machine by a crocodile clip to detect the ST elevation that occurs with myocardial injury. If the procedure is carried out as an emergency in the A&E department or on an ordinary ward the patient should be connected to a cardiac monitor.

Angle of needle The needle should be advanced at 45 degrees to the skin, aimed at the left shoulder, aspirating all the time. A doctor, other than the one performing the procedure, should watch the cardiac monitor or ECG machine for the ST elevation that occurs if the needle touches the heart. If this occurs withdraw the needle slightly.

What do I do with the fluid? It is said that heavily blood-stained pericardial fluid can be differentiated from intra-cardiac blood as it does not clot.

Unless there is an obvious traumatic or iatrogenic cause for the effusion, send off any fluid obtained for protein, glucose, microscopy, culture, and sensitivity including acid-fast bacilli, cytology, and immunology.

What do I do now?   Aspirate as much fluid as you can without the injury pattern appearing on the ECG. The fluid may re-accumulate so if you used a catheter-over-needle or a Seldinger-type central-line or pericardiocentesis set leave the catheter in position, capped off. If a needle alone was used remove it.

If the tamponade was traumatic or iatrogenic in origin refer to the cardiac surgery team urgently. If it was 'medical' tamponade the patient needs further investigation including echocardiography, and cardiac monitoring on a CCU or ITU, as the tamponade may recur.

Further reading

1. Roberts, J. R. and Hedges, J. R. (1991). *Clinical procedures in emergency medicine.* W. B. Saunders, London.

Index